WITNESS TO ROSWELL
75TH ANNIVERSARY EDITION

WITNESS TO ROSWELL

75TH ANNIVERSARY EDITION

Unmasking the Government's Biggest Cover-Up

THOMAS J. CAREY AND DONALD R. SCHMITT
FOREWORD BY EDGAR MITCHELL
AFTERWORD BY GEORGE NOORY

NEW PAGE

This edition first published in 2022 by New Page Books, an imprint of

Red Wheel/Weiser, LLC
With offices at:
65 Parker Street, Suite 7
Newburyport, MA 01950
www.redwheelweiser.com

ISBN: 978-1-63748-003-8

Library of Congress Cataloging-in-Publication Data available upon request.

Cover design by Kathryn Sky-Peck
Interior by Happenstance Type-O-Rama
Typeset in Adobe Caslon Pro, ITC Franklin Gothic, Humanist 521

Printed in the United States of America
IBI

10 9 8 7 6 5 4 3 2 1

. . . [We want] just the facts, ma'am.

—SGT. JOE FRIDAY, Detective Division, LAPD

. . . Record enough facts, and the answer will fall to you like a ripe fruit.

—FRANZ BOAZ, American anthropologist

To my loving wife of fifty-four years, Doreen, not only for suggesting the title of this book, but also for believing in and encouraging me to pursue my "hobby" all these years.

—TJC

To my loving wife, Marie, who inspires me to go beyond the second star to the right, and then straight on 'til morning.

—DRS

CONTENTS

FOREWORD
BY DR. EDGAR MITCHELL

I grew up and went to school in the Pecos Valley of eastern New Mexico. I attended elementary school in the small town of Roswell, and high school in the even smaller town of Artesia, thirty-five miles to the south.

The Pecos River winds its way south from the Sangre de Cristo Mountains of southern Colorado, down the eastern side of the state of New Mexico to eventually join the Rio Grande River, which flows on the western side of the state toward El Paso, Texas. The Rio Grande forms the Texas border between the United States and Mexico. This area is rich with tales of the old West— tales of pioneers, cattle ranches, cattle rustlers, and Indian lore. The small towns and fertile farmlands in the Pecos Valley around Roswell, Artesia, and Carlsbad were part of the wild Western lore surrounding Billy the Kid, Sheriff Pat Garrett, and Judge Roy Bean long before the Roswell Incident, which is the subject of this book.

My local family at that time consisted of my mother, father, and two younger siblings, plus my paternal grandparents, two uncles, an aunt, and their respective spouses and children. In today's vernacular, we were in the agribusiness: farming, ranching, and buying and selling cattle and farm machinery. The head of our clan, my paternal grandfather, was a traditional 19th-century cattleman entrepreneur, known far and wide in the area as Bull Mitchell from his primary interest in bringing registered Hereford breeding stock into the West to replace the longhorn cattle that preceded them. My father and uncles managed the two ranches, two farms, and two farm-machinery dealerships the family acquired in the decade following settling in Roswell in 1935.

I was ready to begin my senior year in high school in the summer of 1947 when the *Roswell Daily Record* on July 8 proclaimed the recovery of a crashed alien spacecraft on a ranch northwest of Roswell. That news caused quite a stir in the small communities in the Pecos Valley, until the following day when the story was retracted after the officials at the local Air Force base declared it was just a crashed weather balloon. The official denial of anything really interesting might have worked had it not been for the close relationships prevailing in the community, such that everyone knew (or knew someone who knew) everyone for fifty miles around. The clamp of secrecy and threats imposed by the Air Force on those immediately involved caused the stories to be told in whispered tones to the closest of confidants, if at all. However, community gossip has a way of persisting and becoming folklore in spite of all efforts to squelch it.

In the immediate aftermath of this incident, I was too busy going to school, working an after-school job, or working on weekends in one of the family enterprises to spend time listening to old folks' stories of crashed alien saucers. However, at family and/or community gatherings sometimes the lore and gossip emerged, particularly if military personnel from the air base, or family of citizenry that had somehow been involved, were present. Then, following high school, for me, the Roswell Incident was only a faint memory from earlier times as I went to college in the East.

In the 1950s, my grandparents passed into the beyond, and the family sold all their holdings in New Mexico and moved on to greener pastures in Oklahoma in the face of persistent drought and declining markets.

Although the events of July 1947 were tucked away in the recesses of my mind during my college years and then during my service with the U.S. Navy as a pilot during the Korean War, they were not totally dismissed. However, it was not until after my mission to the moon on Apollo 14 in February 1971 that incidents related to those early events began to occur.

Although I no longer had family in Roswell or the Pecos Valley, friends from school

Apollo 14 astronaut Dr. Edgar Mitchell in a 2008 photo. (Photo courtesy of Tom Carey.)

and children of family acquaintances were still living there. I had occasion to return to the area for public appearances and talks relative to my adventures in space. I was honored that the highway going south from Roswell to Carlsbad, New Mexico (and that passed through my hometown of Artesia), was renamed the "Edgar Mitchell Highway" in commemoration of my space experience.

During a few of these visits I encountered or was approached by individuals whom I refer to as "old-timers." Also, I was sometimes approached by their relatives of my age, who wanted to discuss the Roswell Incident. Although the incident was almost forgotten in my mind, it was not forgotten in theirs because of harsh threats to their parents or families should the real events ever be discussed. Only because of my being a "spaceman" and a local boy, who in their eyes could be trusted with the secret, did they open up to me a bit. They did not want to carry to the grave what they had observed and been a party to personally or had been told by their parents.

During this period I also met Jesse Marcel Jr., the son of Major Jesse Marcel, the intelligence officer from the Roswell air base who had been on the scene of the crash and brought some of the material home for viewing by his wife and son before taking it to the base. Another Air Force major, an administrative officer who was a friend of my family in those earlier days, whom I shall call "Bill," later verified for me, before he passed on, that he knew Marcel and also knew about the events of the period but was not officially involved in the investigation.

These individuals, plus others related to the Roswell funeral home, the sheriff's office, and the press and radio media at the time, made passing reference to their knowledge of the reality of the Roswell Incident at various social events I attended in that period of the early 1970s.

It was not until many years later, when Stephen Greer and other investigators of UFO events began making a concerted effort for disclosure of alien presence, that I began to think more deeply about these early events and conversations. In 1997, Greer, myself, and Commander Will Miller, while attending a conference in Washington, DC, on this disclosure issue, made an appointment to speak with high-level intelligence officers at the Pentagon regarding our knowledge and experiences. We were granted the interview and told our respective stories to these officers. Although we had adequate confirmation from officialdom following this meeting that our experiences were valid and that there had been an ongoing investigation and official denial of alien presence, even this meeting at the Pentagon and its aftermath were subject to the usual obfuscation and denial. This, of course, is because even the existence of these "special access programs" cannot be admitted by those with the necessary security clearance without jeopardizing that clearance and their jobs.

During the past three decades, well-credentialed and competent investigators, particularly the authors of this volume, have made a compelling case, not only for the Roswell Incident, but also alien visitation in general. For those with an open mind to this idea and willingness to dig through the available literature, the evidence is overwhelming. The question of our being alone in the universe has been a subject of discussion for many centuries, but only in our times has modern technology enabled us to meaningfully enter the discussion with the expectation of finding answers without necessarily experiencing on Earth an alien presence. The visitation phenomenon, as it becomes more widely accepted and understood, will cause a significant change in the view of our place in the universe and a deeper interest in astronomy and cosmology.

The question of traversing vast distances in the universe within finite time periods is a thorny one. However, our science is still very young and incomplete, and our visitors clearly have advanced more in that realm of science than we have. We do have a lot to learn about traveling among the stars, an adventure we have only begun during our generation.

The authors of this book, Thomas Carey and Donald Schmitt, have spent many years ferreting out details of the true story of the Roswell Incident from among the denials, misinformation, and disinformation promulgated by sources desiring to discredit the alien presence. Were a weather balloon (or any other real event) the actual basis of the Roswell Incident, one truthful story would suffice. However, over the years, a host of different cover stories have emerged from officialdom. That fact alone is compelling evidence that the official denial of alien presence for more than sixty years is false. Many credentialed and skilled UFO investigators have told their stories and written articles and books on their investigations around the world. Taken together, they make a most compelling case for this issue. However, the modern era of UFO investigation began with the Roswell Incident. Carey and Schmitt have done a marvelous job of presenting the evidence for the crash at Roswell in exquisitely written detail.

—EDGAR D. MITCHELL
January 2009

PREFACE

How do we define a mystery? To the good people of New Mexico and many others involved in what we have termed "the ultimate cold case file," it can be defined in a single word: *Roswell*. At the dawn of the 21st century, *Roswell* has become synonymous with one of the most important events of all time. For that fact alone, it deserves to be researched and investigated until there is nothing left to investigate, or until a final conclusion is reached that is acceptable to most reasonable minds. The authors believe that the latter option has already been achieved.

To demonstrate this fact, we will build a case for you, focusing on the legal framework and parameters by which the case must be judged. In its way, Roswell has proven to be as painstaking a case to develop and present—spanning now seventy-five years—as any court case that has ever been decided. It is the case for an amazing event and the extreme measures the military authorities took to suppress it. As with any jury, your time and attention are needed to follow the witnesses' evidence. You might find it difficult at times, but it will become evident that another word summarizes this entire event: *cover-up*.

Understanding the case requires us to understand the times in which it occurred. We have to return to an America just two years after victory in her greatest war, when the military was held in perhaps the highest esteem ever by its citizens. We have to return to a time when the predicted post-war depression wasn't happening after all; when the last shreds of Eastern Europe's independence were being torn away with the descent of the "Iron Curtain," and the Cold War was becoming really chilly; when a single Paris fashion designer would decree all hemlines down; a time before air-conditioning (except in movie theaters), when city dwellers slept in parks to escape the summer heat; and when the press had a real tradition of filling the hot-weather news stories with hopes, fads, and wonders. It was a time before television, when radio was the chief means of in-home news and entertainment. Roy Rogers married Dale Evans, and the Brooklyn Dodgers' Jackie Robinson had just broken Major League Baseball's color barrier. The Big Band Era was coming to an end, but rock 'n' roll was still years away, and popular music reflected the uncertain but hopeful outlook of a victorious nation, as charted in *Billboard*'s Top 10 songs for the week of July 6, 1947:[1]

1. "Chi Baba, Chi Baba"—Perry Como
2. "Peg O' My Heart"—Jerry Murad & the Harmonicats
3. "I Wonder, I Wonder, I Wonder"—Eddy Howard
4. "Peg O' My Heart"—The Three Suns
5. "Temptation"—Red Ingle & His Natural Seven w/vocal by Jo Stafford
6. "Peg O' My Heart"—Art Lund
7. "That's My Desire"—Sammy Kay Orchestra w/vocal by Don Cornell
8. "Across the Alley from the Alamo"—The Mills Brothers
8. (Tie) "I Wonder, I Wonder, I Wonder"—Guy Lombardo & His Royal Canadians w/vocals by Don Rodney and The Lombardo Trio
9. "Peg O' My Heart"—Buddy Clark
10. "That's My Desire"—Frankie Laine
10. (Tie) "Peg O' My Heart"—Clark Dennis

The motion picture industry was also in transition from its "Hollywood Goes to War" footing to a post-war "peacetime," however uncertain it may have been. Its patriotic and propagandist offerings during World War II were all but gone, as new terms such as *Iron Curtain* and *Cold War* were becoming part of our lexicon. The top ten films of 1947 appear to reflect a curious pessimism regarding the human condition, perhaps based upon the fact that one war had just been concluded, but the dark clouds of another seemed to be gathering.[2]

1. "Black Narcissus" [Archers]—Deborah Kerr, David Farrar, Sabu
2. "Out of the Past" [RKO]—Robert Mitchum, Kirk Douglas, Jane Greer
3. "The Lady from Shanghai" [Columbia]—Rita Hayworth, Orson Welles

4. "Monsier Verdoux" [Charles Chaplin]—Charlie Chaplin, Martha Raye

5. "Odd Man Out" [Two Cities]—James Mason, Kathleen Ryan

6. "Crossfire" [RKO]—Robert Young, Robert Mitchum, Robert Ryan

7. "T-Men" [Edward Small]—Dennis O'Keefe, Mary Meade

8. "Born to Kill" [RKO]—Claire Trevor, Walter Slezak

9. "Dark Passage" [Warner Brothers]—Humphrey Bogart, Lauren Bacall

10. "The Bachelor and the Bobby-Soxer" [RKO]—Cary Grant, Shirley Temple

10. (Tie) "Miracle on 34th Street" [20th Century Fox]—Maureen O'Hara, John Payne, Edmund Gwenn, Natalie Wood

During the two weeks encompassing the last week of June and the first week of July 1947, newspapers across the country carried accounts describing the arrival of *flying saucers*. Witnesses throughout the country would describe flying "discs" and other assorted, metallic flying objects that defied conventional explanation. Military pilots were placed on twenty-four-hour alert, and radar operators were on twenty-four-hour standby—all looking skyward and hoping that whatever was invading our airspace was not a new threat to our national security that might lead to another war.

The state of New Mexico in 1947 was the most sensitive and highly guarded area in our country, if not the entire world. Not only was there ongoing atomic research at Los Alamos where the first atomic bomb was developed, but there was also the testing of captured German V-2 rockets taking place just to the south at White Sands near Alamogordo. Not far from Alamogordo was also Trinity Site, where the world's first atomic bomb was detonated. And at Roswell itself was the headquarters of the 509th Bomb Group, the only atomic strike force in the world at the time. It was the 509th that, just two years earlier, had dropped the two atomic bombs on Hiroshima and Nagasaki to end World War II. Little did they know that they would also become involved in one of the most significant and historic events of all time: the crash of an unknown object from another world.

In the late evening of July 3, 1947, a severe thunder and lightning storm raked central New Mexico. During the height of the storm, local ranchers would later describe hearing a loud explosion that did not sound like the other thunderclaps. Civilians would arrive at the site first. Some would attempt to report it to the local sheriff. Others would later describe what they saw, but they would wait many years before finally admitting to their closest family members and friends facts that still defy all reasonable and conventional explanation. As you will read, these people, members of America's "Greatest Generation," believed that they witnessed, up close and personal, the remains of an interplanetary vehicle of unknown origin—a crashed *flying saucer*.

ACKNOWLEDGMENTS

T his work is intended to inform those among you, the interested public, who desire to know the truth behind a truly extraordinary event that occurred seventy-five years ago as of this writing. A cover-up of the true nature of this event by elements of the United States military was immediately instituted, and still survives, leaking but essentially intact, to this very day.

The dedicated, proactive Roswell investigative team of Tom Carey and Don Schmitt remains committed to ferreting out and reporting to you the ultimate truth of the so-called Roswell Incident by developing and following every clue, lead, or hint, no matter how small or where it may take us. To that end, this work represents the third major publication of the Carey/Schmitt investigative team—still representing a down payment on history—since it was formed twenty-four years ago and will offer up yet more new information regarding the Roswell events of 1947 resulting from our still-continuing, intensive investigation into this remarkable case of apparent alien visitation.

We would like to thank all of the witnesses, as well as others with source information, who have come forward, many with great reluctance and some with fear for their own well-being, and have agreed to talk to us "on the record" about those long-ago events. Regrettably, because Father Time waits for no one, many of these have since passed away. Without their courage and cooperation, we could not have moved this case forward beyond previous works on the subject, and this publication would not have been possible. Special thanks must go to the late Julie Shuster, herself the daughter of a key player in the 1947 Roswell events, as well as her staff at the International UFO Museum and Research Center in Roswell, New Mexico, for their enduring assistance and support. The IUFOM&RC still remains our base of operations whenever we are in Roswell. Specific thanks must also go to the late Roswell photographer Jack Rodden, who has provided us with several key leads, and who has helped us gain access to especially reluctant witnesses; and to Roswell archaeologist Pat Flanary, whose knowledge of the Roswell region and surrounding terrain has been of immense help, especially in our search for the final Roswell crash site. We also thank:

John LeMay and Elvis Fleming of the Historical Society for Southeast New Mexico in Roswell for their kind assistance in locating relevant photographic images of life in Roswell and the RAAF during the 1940s.

Michael Schratt, for supplying our investigation with several new leads, and for connecting us with John MacNeill (whose computer simulation appears in Chapter 11), to whom we also express our appreciation.

Gloria Hawker, for her assistance in facilitating interviews for the authors with reluctant witnesses, especially Eleazar Benavidez.

The late Earl Fulford, a former sergeant at the RAAF in 1947 as well as a firsthand participatory witness to the UFO recovery operation, who got to fulfill a life's dream of finally talking about it and returning to Roswell where it all happened.

Dr. David Rudiak, for his outstanding work deciphering the "Ramey memo," and for his assistance in identifying some of the early statements from key players that appeared in the national newspaper reports after the initial press release.

Anthony Bragalia, for sharing the ripe fruits of his own independent investigation of the Roswell Incident, especially his interviews with several new witnesses, previously unknown to us, as well as his groundbreaking work regarding the connection between the Roswell physical wreckage, the Air Force, and the Battelle Institute.

INTRODUCTION
BY DONALD R. SCHMITT

Although we are approaching the 75th anniversary of the seminal UFO case, one constant remains; To quote the late, great radio and television broadcaster Larry King, "Roswell comes back stronger every year." Isn't that just the way of that rare commodity known as "truth"? No matter how often you cover it up or attempt to bury it, you can never fully extinguish it. Anything built on truth has an almost indestructible foundation whereas false creations tend to collapse from lack of consistent support. So, King was right in that after four official explanations from Washington and countless attempts by the media to toss Roswell on the ash heap of forgotten history, it refuses to die. Is it merely because we tenacious Roswell investigative authors have made such a sound case or is it more likely that an event this unprecedented stands on its own volition?

Many of our colleagues have joined the skeptical ranks and lament that a story of this magnitude would certainly be next to impossible to hide. Well, it hasn't been hidden—it has been steadily leaking out for over forty years. They just refuse to acknowledge that fact. Those who refuse to look at the evidence will never acknowledge it exists. Others suggest the case is ancient history as all the witnesses have now passed on. But if we are to accept the massive amount of testimony that suggests that a craft and bodies of unknown origin were indeed recovered, such secluded evidence will outlast all of us. No matter where all such proof is hidden or crated away by those who would prevent the truth from seeing the light of day . . . it still exists.

Something extraordinary crashed, was recovered, and then was sent on for testing and analysis—that is a fundamental historical fact. And, as long as that truth still exists, Roswell will never fade away. For to allow that would be the greatest tragedy in all of recorded history. What happened in Roswell is true and we allowed a handful of shortsighted bureaucrats to hide it. But truth has a way of outlasting those who would suppress it. They have not won. Roswell will prevail. The witnesses in this book have seen to that.

—DONALD R. SCHMITT

INTRODUCTION
BY THOMAS J. CAREY

"This is the case. This is the *only* case." So said Boston attorney Frank Galvin, played by actor Paul Newman in the 1982 feature film *The Verdict*. He had said this to his investigator Mickey Morrissey, played by character actor Jack Warden, when Morrissey suggested that Galvin drop the case and just accept the $200,000 offer to settle the case made by the defendant's lawyer Edward Concannon, played by veteran actor James Mason. "There will be other cases, Frank," Morrissey implored Galvin. Galvin had refused the offer on an "epiphany moment" he suddenly experienced when the offer had been made, that if he accepted the offer, then no one would know the *truth* of what had caused his client to become a veritable "vegetable" for the rest of her life. Money would change hands; everybody would go home happy, and that would be the end of it. "OK. What's the next case?"

So it was with me and the Roswell case. I had been a fan of Donald Keyhoe, who had been the chief advocate making the case for "UFOs are real" from the late 1940s through the early 1960s. His books held my fascination of the subject and were well-written and easy to read—even for me, not a big reader of books at the time. One aspect of the UFO phenomenon that Keyhoe would not go near, however, was that of crashed saucers and their alien occupants, living or dead. For Keyhoe, it was a "bridge too far"—too far-out to bring credibility to the subject that was dear to his heart. So, he simply ignored such stories, interesting as some of them were.

In the fall of 1974, there was a major story making its way into the national news cycle about a fellow in Florida by the name of Robert Spencer Carr who was claiming that there were UFO artifacts and alien bodies being stored in a place called "Hangar 18" at Wright-Patterson Air Force Base in Dayton,

Ohio. I wanted to hear more details about it, but the story passed as quickly as it had arrived.

It was in 1978 that I first heard the word "Roswell" spoken. It was by a Canadian UFO researcher by the name of Larry Fenwick, who mentioned that something big was coming about a UFO crash that happened near Roswell. I had no idea where Roswell was located, but I filed it away for future reference.

Two years later, in 1980, I, my wife, and our two children left my hometown of Philadelphia and moved to Huntingdon Valley, Pennsylvania, where we have lived ever since. I also picked up and read a book titled *The Roswell Incident* that told of an alleged crash of a "flying saucer" in New Mexico back in 1947. This was not your "lights-in-the-sky" or "strange-markings-on-the-ground" stuff. It told of the crash of a "nut-and-bolts" craft with crew from another world that had been covered up by the United States Government. The story had all the elements of a good mystery—an otherworldly event first reported as such; then a coverup, including death threats to witnesses, that held for thirty years until one of the key witnesses broke the silence barrier. The book and the story contained in its pages just blew me away! From that point on, all other UFO accounts, past, present and future (so far) have paled into insignificance in my mind when compared to it. Since 1980, the year of my "Roswell Epiphany," it has become the most thoroughly researched and documented UFO case of all time—the so-called "Granddaddy of all UFO cases."

My direct involvement in the Roswell investigation took place in 1991, when I joined the investigative team of Kevin Randle and Don Schmitt to try to locate the archaeologists that were mentioned in *The Roswell Incident* book as having discovered the downed spaceship. That investigation took me two years to complete, and as a result we were able expose a major hoaxer and also eliminate the Plains of San Agustin west of Socorro from the Roswell story as being the location of the crashed saucer with dead aliens.

I took my first trip to Roswell in 1993 to meet "the boys" in person for the first time, after which I figured that was pretty much it—the end—of my active involvement in the investigation of the Roswell case. After the 50th Anniversary-year hoopla of the Roswell crash in 1997, most of the prior investigators of the case left the field thinking there was essentially nothing more to investigate. Don Schmitt and I felt differently. So, we teamed up in 1998 to continue a *proactive* investigation of the Roswell case that resulted in the publication of the first edition of this book.

Even now, on the 75th anniversary of the incident, our investigation continues. With hundreds of additional first- and secondhand witnesses located and interviewed by us, we believe that we have incontrovertibly made the case for Roswell as being one of an "extraterrestrial visitation by a craft and crew of unknown origin."

A number of lawyers who have read our books have written to us to tell us that if we took our case to court, hypothetically against the U.S. Air Force, the Defense Department, or the U.S. Government, we would win the trial "hands down!" And when people want to know why I am *still* investigating a seventy-five-year-old case, I simply tell them it's because, "This is the case. This is the *only* case."

—THOMAS J. Carey

1

THE ULTIMATE COLD CASE FILE

Crime shows have proven to be a popular and hardy staple for TV producers and viewers alike for decades. Shows—such as *Man Against Crime, Gangbusters,* and *Dragnet* from the early days of TV, through the more sophisticated *Peter Gunn; Richard Diamond, Private Detective;* and *Perry Mason* of the late 1950s and 1960s, to *Kojak, Columbo,* and *Hill Street Blues,* right down to today's ultra-legalistic *Law and Order* and super high-tech *CSI: Crime Scene Investigation*—though separated in time culturally, stylistically, and technologically, all share one common theme: **the search for truth.**

In early July of 1947, *something* crashed to Earth in the high desert (higher than 2,000 feet) of eastern New Mexico during one of those severe thunder and lightning storms that occur in the region every year during monsoon season. A few days later, the U.S. Army Air Forces (as the U.S. Air Force was called until later that year) electrified a nation and the world by issuing a press release announcing that its 509th Bomb Group at the Roswell Army Air Field (RAAF), located just south of the sleepy New Mexico town of Roswell, had "captured" a flying saucer that had crashed nearby. Within hours, however, a press conference was hastily convened at the Eighth Air Force Headquarters in Fort Worth, Texas (the command to which the 509th Bomb Group was attached), to announce that it was all a big mistake. The *flying saucer* was nothing more than a misidentified *weather balloon!* The press immediately lost interest, and the story quickly died. Outside of the occasional rumor, the story was then forgotten and remained buried for the next thirty years. Then, in 1978, the intelligence officer of the 509th Bomb Group at the time of the incident broke the silence by publicly stating that what crashed outside of Roswell in 1947 was no weather balloon, but something "not of this Earth." A few interested UFO investigators took note and undertook a civilian investigation of the case. By the early 1990s, several books had been written on the subject, all favoring an extraterrestrial conclusion, and the case was prominently featured on the popular TV show *Unsolved Mysteries*, which also strongly suggested an extraterrestrial answer for the mystery. As public awareness of the case grew, pressure mounted for some form of official restatement by our government concerning its position on the matter. This occurred in 1994 when the Air Force admitted that it had in fact lied in 1947 with its weather balloon explanation, but it was now telling the truth with its *third* explanation for Roswell: what crashed was now a high-flying contraption composed of multiple balloons, multiple radar targets, and a listening device belonging to a special project—Project Mogul—that fell to Earth near Roswell. Although the project's purpose—to detect sound waves from the anticipated detonation of the Soviet Union's first atomic bomb by employing high-altitude, balloon-borne, acoustic sensors—was top secret, its off-the-shelf components were not. Far from it—the prosaic rubber balloons, tinfoil radar targets, and balsa wood struts used in the project were the *exact same types* used in most weather balloons and radar targets of the time—materials that any six-year-old would have no trouble identifying! Then, in 1997—the 50th anniversary of the Roswell crash—the Air Force brazenly offered up its *fourth* explanation, this one to try to deal with the long-held rumors and eyewitness accounts of diminutive (3½- to 4-foot-tall) "alien bodies" that were also alleged to have been recovered from the Roswell crash. Known as the "dummy explanation" for obvious reasons, the Air Force spokesman was met with derisive howls of laughter from members of the press when he attributed such claims to the Air Force's use of full-size (6-foot-tall) mannequins in several projects involving high-altitude parachute drops that were conducted in New Mexico *in the 1950s* in preparation for our country's manned space program. To explain away the ten-year time disparity, the Air Force claimed that the witnesses were unwitting victims of a mental processing affliction known as "time compression," whereby recollections of past events tend to contract the time frames in which they took place as a person ages. Thus, those who claimed to have seen alien bodies from the Roswell crash in 1947 were really remembering a chance encounter with crash-test dummies that they somehow stumbled upon while searching for rattlesnakes out in the desert in 1959! Project Mogul and "dummies-from-above" continue to be the Air Force's "explanation" for the Roswell crash.

The search for truth in the real world of trying to solve cases ideally involves a twofold investigation of pertinent facts: (1) the search for incriminating, physical evidence, from old-fashioned fingerprints on the murder weapon to the currently trendy and "infallible" DNA evidence at the crime scene (in other words, "forensics"), coupled with (2) relevant and credible eyewitness testimony. When presented to a jury in a logical and coherent manner, this investigative combination constitutes "proof" as we know it and has, with notable exceptions, stood the test of time in proving to be a case-winner. In the absence of direct physical evidence, the prosecution will attempt to build a circumstantial case against a defendant based solely upon witness testimony as the proof—yes, witness testimony in and of itself is considered evidence, and if sufficiently convincing, *proof*—in every courtroom across the land, and this testimony has indeed sent many a defendant to the electric chair. This fact must be stated up front because of constant complaints by skeptics and debunkers of the so-called Roswell Incident[1] who appear not to know or respect the legal standing of witness testimony, by downplaying it or ignoring it altogether when it comes to the subject of UFOs, and especially Roswell. Regarding the latter, the standard line goes that because there exists no incontrovertible *physical evidence* to support the case for an alleged crash of a UFO near the town of Roswell, New Mexico, in 1947, it did not happen. Case closed.

Unfortunately, not every criminal or civil investigation results in an outcome involving a prosecution or other resolution, often due to a lack of evidence to move the case forward to an indictment or charge. When this impasse persists throughout a period of time, a case becomes what is euphemistically known as a "cold case file." Today, there are thousands of such cases stuffing overburdened police file cabinets in every jurisdiction, most of which unfortunately will never be solved.

A recently popular TV show, *Cold Case Files*, has taken up the theme of trying to solve old, seemingly unsolvable cases that have languished for years. Although these are fictionalized accounts, many of the show's weekly offerings are based upon real-life cases that have lain dormant, on the average, for ten or twenty years. What we find interesting about this show are the parallels between the investigative tools employed by the show's Cold Case Investigative Unit in trying to investigate an aged case, and our experience in investigating the Roswell case. The parameters are remarkably similar, but with a few notable differences. The *Cold Case* investigators still have at their disposal as potential targets of investigation a combination of still-available, fertile sources of case evidence such as: (1) physical evidence just waiting to be uncovered and subjected to forensics, (2) key, finger-pointing documents readily available on microfiche at local repositories, and especially (3) still-living, easily located witnesses willing, however reluctantly, to belatedly spill their guts. The happy result of all this activity is usually another cold case file scoped, investigated, and solved—all in one hour's time!

Considering the fact that we have a combined total of forty years of investigative experience dedicated solely to uncovering the truth of the Roswell Incident, and the fact that the case still remains unsolved, or at least unproven in a majority of the public's minds more than seventy-five years after the event, we see the Roswell Incident as the ultimate cold case file. Most of the original investigators who were once active on the case during the 1980s and 1990s have left the field and will engage again only if something worth their additional effort drops into their laps. We remain the only proactive investigative researchers still working the Roswell case.

Although there have been several pieces of metal submitted to various UFO investigators through the years purported to have come from the Roswell flying saucer, upon analysis none were found to be sufficiently exotic in terms of their constituent elements or construction to be considered unequivocally as coming from another world. In one particular instance, the subject piece of metal was indeed unique, but upon inspection turned out to be nothing more exotic than a piece of Japanese jewelry! Meanwhile, our search for physical evidence—the Holy Grail of the Roswell case—continues.

Document-type evidence in the Roswell case consists mostly of the newspaper accounts from 1947 announcing the "capture" of a flying saucer near Roswell, followed quickly by a retraction of that story in the guise of a "misidentified weather balloon" that stood as the explanation for the next thirty years. Because of the timing of the original press releases, newspapers in the eastern United States carried both versions of the story on the same day (July 9, 1947), whereas most newspapers in the western time zones carried the flying saucer story one day (July 8, 1947) and the weather balloon retraction story on the following day (July 9, 1947). At a minimum, the newspaper accounts verified the fact that something did in fact crash near Roswell in early July of 1947, and key participants in the event were named. Just what crashed became (and still is) the issue.

Since the mid-1980s, various government documents, alleged by some Roswell investigators to be genuine, have surfaced that would seem to verify the truth of the crash and recovery of an extraterrestrial spaceship and its crew in southeastern New Mexico in 1947. Referred to collectively as the MJ-12 Documents, they highlight some of the problems associated with "documents-as-proof" that are frequently encountered in UFO-related research: lack of provenance for the document(s) in question, and a lack of agreement among researchers regarding the genuineness of a document. Tracing a document back to a UFO investigator with an agenda and no farther does not constitute adequate provenance. We must know who the ultimate originator of the document in question was in order to properly research and verify its context and background. Without its provenance thus established, a document should immediately become suspect as being fraudulent, especially if the original of the document is unavailable. To make matters worse, UFO researchers cannot agree on the importance that should be attached to apparent discrepancies when they are found in such documents. The finder of the discrepancy—whether in the form of a incorrect date format, a misplaced comma, or a questionable signature—will discount the entire document as fraudulent, whereas the advocates of the document will attempt to downplay the discrepancies as minor matters signifying nothing, in order to save its hoped-for document-as-proof status. The point here is that without standing and general agreement regarding its import, a document as proof of anything is worthless.

In 1993, the Government Accounting Office (GAO), which is the investigative arm of Congress, at the request of the late New Mexico Congressman Steven Schiff, undertook a search of all relevant government agencies (the Department of Defense, the CIA, the Air Force, and so on) for documents relating to the 1947 Roswell Incident, for the purpose of establishing a paper trail of events from the appropriate time period. The results of the search were published by the GAO in 1995 and, instead of clarifying things, served only to muddy the waters even more.[2] No additional documents were turned up by any of the agencies tasked by the GAO beyond those very few documents that were already known. But a bigger surprise came when it was discovered that all documents, such as teletype messages, telexes, radiograms, letters, invoices, and other records emanating from the Roswell Army Air Field (as the late Walker Air Force Base, which closed in 1967, was known in 1947) covering the general time frame of the Roswell Incident had been destroyed years before *without explanation or apparent authority*. To Roswell investigators such as ourselves, already convinced of a massive government cover-up of this case, there can be no innocent explanation for this. To skeptics of the Roswell Incident . . . well, it was just one of those things signifying nothing.

One final introductory note regarding documents as they relate to the Roswell investigation has to do with two documents whose provenance is not an issue. The first is an FBI memo dated July 8, 1947, that was written by the FBI's agent in Dallas, Texas, to their office in Cincinnati, Ohio, shortly after the Eighth Air Force's commanding officer, General Roger Ramey, held a press conference in his Fort Worth office and announced to the world that the flying saucer recovered at Roswell a few days earlier was nothing more than a misidentified weather balloon. The memo suggested that a lie had been perpetrated upon the public at the staged press conference, and that the flight transporting the Roswell wreckage from Fort Worth Army Air Field (where it stopped after leaving the RAAF) to Wright Field in Dayton, Ohio, had not been cancelled, as General Ramey had dramatically declared to the press. The second document is referred to by many as the "smoking gun" of the Roswell case. It is a telex that was held in the hand of the architect of the Roswell cover-up himself, General Ramey, during his aforementioned press conference, which seems to tell a different story from the one he was giving to the press.

Several key witnesses to the Roswell events of 1947 unfortunately passed away long before the case appeared on the radar of UFO investigators. The sheriff of Chaves County, George M. Wilcox, for which Roswell is the county seat and who was used by our military to assist them in the cover-up, passed away in 1961; the Corona sheep rancher, William W. "Mack" Brazel, who first discovered his pasture filled with strange wreckage—and something else— and thereby was responsible for starting the entire Roswell chain of events, died in 1963, the same year as General Ramey; the commanding officer of the RAAF in 1947, Colonel William H. Blanchard, died of a heart attack at his desk in the Pentagon in 1966 as a four-star general and vice chief of staff of the Air Force; and Major Jesse A. Marcel, the intelligence officer under Colonel Blanchard at the Roswell base in 1947, passed away in 1986, but not before breaking his thirty-year silence in 1978

Grave headstone of Roswell hardware store owner Dan Wilmot, who reported seeing a flying saucer from his front porch heading northwest in the sky late on the evening of July 2, 1947. (Photo courtesy Tom Carey.)

regarding an event to which he bore witness and believed to his dying day originated "not from this Earth," thereby igniting a controversy that is still with us today.

The Department of Veterans' Affairs estimates that veterans of World War II, whose numbers include most of those involved in the Roswell Incident, are leaving us at a rate of about 1,500 per day. And, as any insurance actuary will tell you, that statistic will swell at an ever-increasing rate with the passage of time. For this reason, we have gone beyond racing the undertaker and are dangerously close to the finish line with our investigation. (It is sobering indeed to note that the youngest known participant in the 1947 Roswell events, a then seven-year-old boy named Dee Proctor, passed away in 2006 at the age of sixty-six.) And of those who are still with us, many have succumbed to the ravages of old age, such as Alzheimer's, Parkinson's, and so on, and cannot now be interviewed. As a result, we find ourselves increasingly interviewing the children and grandchildren of the actual participants, and an alarming number of *this* group is starting to pass on as well. We estimate that, of the military personnel who were stationed in Roswell in 1947, *at least* 90 percent are now unavailable to us because of death or infirmity. These same statistics also hold true for the civilian population living in and around Roswell and Corona in 1947. It is due to the complete lack of verifiable physical evidence, the dearth of acceptable documentation, and mostly to the ever-decreasing nature of the witness pool after the passage of seventy-five years that we refer to Roswell as the ultimate cold case file.

Although we believe that crimes were committed by our military against civilians in Roswell and Corona during its heavy-handed suppression and cover-up of the Roswell Incident, no charges were brought at the time, and the statute of limitations has long ago run out. The case still remains, however, an active historical mystery left to us from the 20th century to solve. Even allowing for the given limitations facing the Roswell investigation, we have amassed what we believe is overwhelming evidence to sustain an extraterrestrial conclusion for the Roswell events of 1947, enough to prevail in any court case against the U.S. Air Force and its balloon explanation. Any such hypothetical court action would involve a simple "preponderance of the evidence" standard of proof (used in civil cases) as opposed to the higher standard of "beyond a reasonable doubt" (used in criminal cases) to render a verdict. The truth be told, the Air Force's case for Roswell is so water thin (there are no physical evidence, pertinent documents, or credible witnesses, living or dead, to connect a balloon event to the Roswell Incident) that it would also be a sure loser in court—beyond a reasonable doubt.

In science, where the bottom line is also the search for truth, Occam's law of parsimony, a.k.a. Occam's Razor, is used to decide among competing hypotheses the one that best explains the observed data. It holds that, with all other factors being equal, the simplest hypothesis that explains the most observed data is the best and must prevail. Thus employed, it serves as a tool for eliminating competing-but-lacking theories, hypotheses, and explanations from others being considered. In the Roswell case, competing hypotheses as to what might have crashed in 1947 (a V-2 rocket, an experimental rocket plane or jet aircraft, a propeller-driven Chance Vought "Flying Flapjack," a Northrup "Flying Wing," a Japanese "Fugo" balloon-bomb, or an errant atomic bomb) have all been investigated and eliminated. That leaves us with two remaining, competing hypotheses: the Air Force's high-altitude Project Mogul and time-compressed memories of anthropomorphic dummies from the sky[3] versus a crash of a bona fide UFO—a flying saucer in 1947 terms— along with its unfortunate crew, which is favored by most civilian researchers. Which is the bogus hypothesis?

Without conclusive physical evidence and/or accepted documents to decide the case one way or the other, the Roswell Incident is of necessity a "witness case." (It should be pointed out, however, that we are continuing a proactive search for all three types of evidence.) On the "anti-Roswell" side of the debate, the Air Force can offer *not a single, credible eyewitness* to a balloon event at any of the three Roswell crash locations identified by us or during the subsequent recovery operation. There is not a shred of evidence—nothing—to connect a wayward Project Mogul balloon-train to what crashed in Mack Brazel's sheep pasture in July of 1947; time-compressed dummies from the sky as the answer for the purported recovery of "little bodies" from the crash is so far-fetched that no one, save a few stolid mainstays of the so-called mainstream media[4] who normally do not accept *anything* that our military says at face value, has taken it seriously—not even Roswell skeptics!

On the "pro-Roswell" side of the debate, we can offer scores of credible eyewitnesses, military and civilian, out of a total witness pool to date of several hundred first-, second-, and thirdhand witnesses to an *extraterrestrial* event that we know today as the Roswell Incident. None of these knows the entire story of what occurred, as each knows only what he or she witnessed or took part in. It has been our task to piece together, like a jigsaw puzzle, these moment-in-time snapshots into a larger picture of what took place so long ago. Because of the wealth of credible witness testimony that we have secured in our investigation of the 1947 Roswell crash (we have witnesses along the entire timeline of events, from the discovery and recovery of the UFO and its crew at the crash sites, their initial transport to the base at Roswell where a preliminary autopsy was attempted, the flight to Fort Worth where the cover-up began, and the flight to their final destination at Wright Field), we have been able to formulate a coherent chronicle for a case of apparent alien visitation. For this reason, applying a healthy dose of Occam's Razor to the known facts of the Roswell Incident *must* result in a rejection of the Air Force's current explanation as bogus.

2

FALLING ON THE AIR FORCE
SWORD . . . FIFTY YEARS LATER

To date, the investigative team of Don Schmitt and Tom Carey has interviewed more than 600 witnesses directly or indirectly associated with the Roswell Incident of 1947. Of these, more than 150 were firsthand witnesses to something truly extraordinary, and many of them, to their very deathbeds, broke decades of duly sworn silence to events that they personally experienced. Man, woman, and child—each and every one of them has described events chronicled in numerous articles and in our best-selling books. Each and every soul brave enough to speak out on the public record against the cover-up of the facts about what truly transpired out in the high desert of New Mexico that summer has reinforced the position that the U.S. military first stated in that press release on Tuesday, July 8, 1947, "We've captured a flying saucer." Those not reluctant to talk all agree with that "official" explanation . . . that is all but one.

Today, former CIC officer and eternally reluctant Roswell principle participant and witness Lt. Colonel Sheridan Cavitt represents the only individual who stood with the Air Force's modern-day efforts to squelch the flying saucer crash. Cavitt's public position is that Roswell merely represents the recovery of a very common weather balloon device. But with Cavitt, the implications are much more serious as he remains a confirmed firsthand witness and happened to be in the position to also receive orders from Washington. This is in stark contrast to his original claims to us: having not been there and total ignorance of what happened in Roswell.

When first approached by Roswell investigators in the early 1980s about his possible connection to the case, Cavitt denied that he ever knew a Major Jesse Marcel or that he was stationed at the RAAF in the summer of 1947. Then, and only when confronted with documentation that demonstrated otherwise, he conceded that, rather, he was on leave for a wedding during the critical time in question. Again, he was confronted with documentation to the contrary that placed him not only at the base but, astonishingly, on the very ranch that the incident took place! And none other than Major Jesse Marcel described the two of them accompanying rancher Brazel to the very spot where he discovered the remains from the crash.[1]

We provide a little background about the Counter Intelligence Corps (CIC) in order to demonstrate the significance of Cavitt first being sent out with Marcel: In 1947, the CIC provided intelligence to both the U.S. army and the Department of Defense (DOD). It was responsible for collecting, producing, and disseminating military and military-related "foreign" intelligence, including intelligence on indications, warnings, capabilities, tactical weapons, and equipment. The question then becomes obvious: Why was the head of CIC at Roswell ordered to accompany the head of intelligence Marcel out to the ranch? The answer is most significant to the entire Roswell affair and seldom addressed: As a CIC officer, Cavitt would have been in the best position to determine if the "unidentifiable" wreckage was of "foreign" design and origin. More evidence was required and the small amount which Brazel brought into town was not sufficient to make that assessment. It becomes abundantly clear that none of the senior officers at the base were able to identify the source of the wreckage.

It has been widely known since the late 1970s, when Jesse Marcel was first interviewed by media and investigators, that a plainclothes CIC officer had accompanied Marcel and Brazel back to the foster ranch on July 6 and 7 of 1947 to inspect the strange debris field that the rancher had reported to Chaves County Sheriff George Wilcox, to radio station KGFL announcer Frank Joyce, and finally to Marcel. Moreover, that man was former CIC Captain Sheridan Cavitt, which was also confirmed by official military documents and corroborated by others stationed at the RAAF in 1947, including Cavitt's immediate subordinate in the CIC unit there, Master Sergeant Lewis B. "Bill" Rickett.

During the course of numerous interviews conducted by researchers throughout the years (prior to his death in 1986 at the age of seventy-eight) Marcel stated that after reporting what he had seen of the Brazel debris in Sheriff Wilcox's office to base commander Blanchard, the colonel ordered Marcel to accompany Brazel back to the Foster ranch in Lincoln County to see exactly what was there, adding that Cavitt should go along with him. The two followed the rancher, Marcel in his own 1942 baby blue Buick convertible, and Cavitt in a jeep carryall that Marcel had quickly checked out of the motor pool.

Upon arriving at the ranch on Sunday evening, July 6, the two officers examined the largest piece of debris that Brazel had stored in a small livestock shed within a short walking distance from where they would spend the night. Since it was approaching dusk and they were still three miles from the debris site, the three spent the night in a small bunkhouse called the Hines House.

At daybreak, Brazel and Cavitt saddled up horses (Cavitt was from San Angelo in West Texas and knew how to ride) while Marcel followed them to the debris field in the Army jeep. Once there, the 509th intelligence officer discovered an area about three-quarters of a mile long by several hundred feet wide covered with, among other items, bit and pieces of thin, weightless, foil-like, metallic debris the color of dulled aluminum with incredible strength and other peculiar properties. His immediate impression was that whatever had caused this scene must have exploded mid-air rather than crashed to the ground. His firsthand account of what the wreckage looked and felt like (confirmed by his son Jesse Jr., to whom he had shown a portion of it in the very early morning of July 8 on his way back to the base) is well known. "Nothing made on this earth"[2] is how Marcel later described what he had found in the grazing pasture that day. Jesse Marcel Jr., eleven years old at the time, stated until his death in 2013 that his father was convinced—even at the time of that early morning investigation of the crash scene—that the debris his father displayed for him and his mother on their kitchen floor was from a flying saucer. At first, Brazel described the debris as "from one of them flying saucers,"[3] but later was coerced into recanting, saying that he "didn't have it quite right the first time."[4] The only other member of the threesome that went out to the debris site that day, Sheridan Cavitt, denied even being there . . . but only when first questioned.

Marcel's participation in the initial recovery effort was documented in the July 8, 1947, edition of the *Roswell Daily Record*, and the same paper confirmed Brazel's involvement one day later. Cavitt, probably because of his counterintelligence role at the RAAF given that his primary duties involved working undercover and searching out Soviet spies, was not mentioned in either of the accounts.

The issue here is not to debate what was found on the J. B. Foster ranch where the aforementioned events took place, but to clearly identify the key players, and more specifically to identify Sheridan Cavitt as one of the military team members, along with Marcel, who accompanied Brazel. Roswell researchers may debate other aspects of the case but all are in agreement (even the Air Force) that there was indeed a debris field where something came down on the Foster ranch about fifteen miles southeast of Corona, New Mexico. The only argument today centers around *what* was found there, not *who* was there.

Cavitt gave an exclusive and extensive interview to the Air Force's Colonel Richard L. Weaver on May 24, 1994, in preparation for the Air Force's response (*Newsweek* magazine called it a preemptive strike) to the anticipated General Accounting Office (GAO) report on Roswell commissioned by the late New Mexico congressman Steven Schiff. He described to Weaver what he claimed was the extent of his involvement in those events in early July 1947. Incredibly, this testimony was given to Weaver at the same time that Cavitt was still denying to us personally that he was even present at the RAAF at the time. For example, when we interviewed him again a month after Colonel Weaver had met with him, Cavitt reverted to his previous story that he was not involved at all. One is compelled to ask if Cavitt's selective memory is affected by who the interviewer is. Did he think that his testimony to Weaver confirming his involvement would be kept secret?

In what amounts to a tacit confession, Cavitt described to Weaver that, yes, he was personally involved in the Roswell events after all, thus officially going on the record admitting that he had been deceiving and misleading Roswell investigators for almost ten years. Readers can draw their own conclusion from the veracity of Cavitt's declaration to Weaver: "I am telling the truth, and I have told all of these other people [Roswell investigators] the truth."[5]

Weaver's interview of Cavitt was no doubt intended to be nonconfrontational and conducted to elicit a specific result. Weaver, who was reading from our first book *UFO Crash at Roswell* (New York: Avon, 1991), comes across as an officer very unfamiliar with the concept of asking any follow-up questions when faced with incomplete, inconsistent, or incomprehensible responses. Instead, he afforded the fellow officer the courtesy of one leading question after the next and even resorted to helping Cavitt at times when faced with providing an accurate answer. Nevertheless, Cavitt seemed wary about being questioned again about Roswell.

For her part, Cavitt's wife Mary chimed in whenever she could to support her husband's story that he really didn't know anything, even though she admitted her husband had never discussed anything related to his intelligence work with her. The result was a paradoxical mix of opaque vagueness or a total memory lapse regarding the Roswell Incident ("It's hard to remember July '47; I hadn't been there very long"[6]), contrasted with details when he recalled other non-Roswell-related events of that time period (for example, Cavitt's description of a 1947 B-29 crash that occurred near Roswell was surprisingly lucid and quite articulate).

Whether deliberately or otherwise, Cavitt claimed not to remember much about Roswell events, except that he told Weaver of finding the immediately recognizable remains of a weather balloon, some aluminum foil, and bamboo sticks on a remote ranch outside of Roswell, all of it quickly gathered up. Unstated and unanswered was the most logical of all logical questions: Why would Cavitt have allowed Colonel Blanchard and Major Marcel to declare to the world that they had recovered a flying disc, only to be made to look foolish later when it was identified by their commanding general, of all people, as nothing more than a common weather balloon device?

Given the day of the week and his description of the site, it is clear that Cavitt was describing for Weaver his trip to the Foster ranch debris field on Sunday, July 6: "We went out to this site. There were no . . . checkpoints or anything like that (going through guards and that sort of garbage) . . . we went out there and found it."[7] The final impact site was much closer to Roswell, at which the remains of the craft and crew had already been reported and subsequently cordoned off with military police by the end of the day on Monday, July 7. Thirty miles to the west, the Foster ranch would not have been secured until early Tuesday morning, July 8. In any event, Cavitt's memory appeared confused when he tried to remember exactly whom he went with and precisely which location. He couldn't seem to decide if he had gone out with Marcel or Rickett, his own CIC subordinate, or both.

Contrary to the record previously discussed, Cavitt told Weaver that he was pretty sure that Rickett was with him, but he wasn't sure about Marcel: " . . . we heard that someone had found some debris out not too far from Roswell and it looked suspicious; it was unidentified. So, I went out and I do not recall whether Marcel went with Rickett and me . . . I had Rickett with me . . . I'm not sure it was Marcel, but I know Rickett was [with me]." Then, switching gears and seeming to go the other way, "I do not remember whether Marcel was there or not on the site. He could have been . . . more and more thinking back on it now he [Marcel] must have been . . . I must have been with him . . . Marcel had gotten a Jeep out of the motor pool."[8]

No doubt growing more confused by all of the bobbing and weaving (no pun intended), Weaver on two occasions during the interview tried to suggest that Cavitt might have made more than one trip to the Foster ranch. The suggestion elicited this response from Cavitt: "I went back down there? No, no . . . I can't recall ever making more than one trip."[9] That would fit nicely with a weather balloon story. Why make another long, hot, dusty drive out into the middle of the desert to observe some downed weather balloon wreckage—which supposedly had already been gathered up and retrieved?

Adding further confusion to the mix, and perhaps trying to have his account of the events track with that of a routine recovery situation, Cavitt told Weaver twice during the interview that he never met Mack Brazel: " . . . I never met the rancher, Brazel or Brazzel, whatever his name was."[10] Perhaps trying to cover all of his bases and solve this curious dilemma of who was where, when, and with whom, Cavitt's signed statement of Weaver's report dated May 24, 1994, states the following:

> Shortly after arriving at Roswell, New Mexico, in that time frame [late June or early July 1947] I had the occasion to accompany one of my subordinates, Master Sergeant Bill Rickett, CIC, and Major Jesse Marcel, Intelligence Officer [of the] 509th Bomb Group, to a ranch land area outside of Roswell to help recover some material. I think that this request may have come directly from Major Marcel.

I do not know who may have made the report to him. To the best of my knowledge, the three of us traveled to the aforementioned ranch land area by ourselves (that is, no other person, civilian or military, were with us). I believe we had a military jeep that Marcel checked out to make this trip.

The undeniable problem with this statement is evident to anyone who has ever made the trip from Roswell to the former Foster ranch and has tried to locate the 1947 debris field site. Without someone to guide you (if you haven't been to the precise location before), it simply cannot be done. From Roswell, the trip today takes over two and one-half hours, miles of it over gravel roads and then across rocky, washed-out, open rangeland. It would have been even more difficult and more time consuming back in 1947. Even today, it is literally impossible to draw up a map directing someone to the site. Without Brazel to lead them, Marcel, Cavitt, and Rickett simply could not have found the way on their own. Is Cavitt's memory just faulty or confused by the passage of time?

It should be pointed out that, at the time of Weaver's interview, Marcel and Rickett were both deceased; therefore, follow-up interviews and possible rebuttals to Cavitt's remarks were not possible. However, both Marcel and Rickett had been interviewed extensively throughout the years, and written as well as a taped record (both audio and video) of their testimony has been available to researchers for some time.

In July 1947, Lewis B. Rickett was a plainclothes master sergeant in the CIC, and both he and his boss Sheridan Cavitt were early 1947 graduates of the CIC training school located at camp Holabird in Baltimore, Maryland. Rickett recalled that he had arrived at Roswell in the spring of 1947 some weeks before Cavitt arrived in mid to late June to take over the CIC unit there. Rickett was sure that "this deal" (as he sometimes referred to the Roswell Incident) had occurred prior to "the split" (the official separation of the Air Force from the Army into two separate branches in September 1947). Rickett later chose the Air Force, where he became a member of the Office of Special Investigations (OSI)—the Air Force equivalent of the Army CIC. His career included a stint as a section chief at the Pentagon, ultimately reaching the retirement rank of Lt. Colonel.

Rickett recalled that "about the time" he had gone out of town (by deductive reasoning, this was on Monday, July 7) to conduct a background investigation on someone outside of Roswell and had returned home late that same evening. The next morning, he reported back to the CIC office around 11 a.m. It was then Tuesday, July 8, and Cavitt wasn't at his desk. Rickett recalled how the CIC secretary explained to him that his boss had left in a big hurry with Major Marcel and "some farmer-looking person . . . they all got some vehicles and took off" and that he had left instructions that he wanted to meet with Rickett around 1 p.m. that same afternoon. According to the noncommissioned officer (NCO), Cavitt returned around 1:15 and immediately told Rickett to take a drive with him, stating, "I don't believe what I've [seen] . . . I want you to go with me somewhere." Rickett, still in the dark at this point as to what was going on, could only respond, "Where you been? You and Marcel went out of here with some old farmer . . . " Cavitt, perhaps demonstrating control of the situation to a subordinate, corrected him on this point ("some old rancher—not farmer"). As to where they were going, " . . . to the boondocks"[11] was all that Cavitt would say.

According to Rickett, he and "Cav" both changed into another pair of shoes that all CIC operatives kept handy for fieldwork, in this case suitable for walking in desert terrain, then drove out to a place that Cavitt wanted to show him. Just the two of them were in the vehicle, an old Plymouth staff car; Cavitt did the driving. Rickett was concerned about getting stuck in the sand, but Cavitt assured him " . . . there's a halfway road out there."

When they arrived at the site, they saw a number of military vehicles and were met by five or six MPs, at least one of whom Rickett recognized as being from the Provost Marshal's 1395 unit command at the RAAF. All were armed with drawn .45s and a few were brandishing submachine guns. Even though they were known by the MPs, their identification was still checked. "They weren't taking any chances," and they were allowed to pass. They proceeded another 100 yards up the two-track dirt trail and ran into the Provost Marshal himself, Major Edwin Easley, and the balance of his men, twenty-five to thirty MPs scattered "here and there, on top of these little dunes," out 300 to 400 yards in a big circle, guarding the perimeter to keep any intruders from wandering into the heavily secured area.[12]

Rickett described the area as mostly flat desert scrub interspersed with rolling knolls or dunes, some high as 15 feet, which he thought was unusual. To the authors, having been to this very location, this description is entirely reminiscent of the final impact site just thirty-five to forty miles north of Roswell, where the remains of the craft and crew were recovered. This location is thirty miles east-southeast of the Brazel debris field, which would have required a drive of over three hours from Roswell back at that time; Rickett distinctly remembered their drive from Roswell was only about forty-five minutes, which is totally consistent with the impact site much closer to town. Rickett also stated that they returned back to the office with sufficient sunlight about 5 p.m. If they had originally left the office for the Foster ranch location at 1:30 p.m., as Rickett described, the earliest that they could have returned back to the base would have been nine-thirty to ten, late that evening, assuming that they did not spend any time surveying that location. If they had spent any time performing more than a cursory examination at that spot, they would have faced a treacherous drive back in the dark from the desert cross-country in a vehicle hardly adept at such an off-road excursion. Such an unlikely scenario would have taken a much longer expanse of time and effort and would have been extremely dangerous. None of this was recounted by Rickett. Clearly, his recollection is accurate; they traveled to a specific destination much closer to Roswell than the Foster ranch—the impact site.

Still, could Bill Rickett have been confused or mistaken? Cavitt had recalled to us in a separate incident, as well as to Colonel Weaver, their finding a balloon-borne radiosonde (radio-signal tracking instrument) near Ruidoso, New Mexico, which is due west of Roswell and would have been a two-hour drive each way from town. Unlike Cavitt's, Rickett's testimony was always consistent and very detailed as to where and what. Such was his intelligence training as a field operative. Cavitt's story changed each and every time he was interviewed. Such is the training of an intelligence officer whose mission is often to skew and distort the data from the field—specifically in matters of national security. It remains ironic that when Cavitt voiced the new "official" explanation of the U.S. Air Force, they latched on to him and not Rickett or the hundreds of other witnesses who endorse the recovery of an actual flying saucer.

While walking the site, Cavitt reiterated his reasoning to Rickett for bringing him there: "I just thought it would be advisable for someone else to see it." Once again demonstrating control and authority over the situation, he told Rickett not to walk anywhere without permission but at the same time asked for his opinion as to what he was observing. Rickett's initial perspectives to Cavitt are intriguing and reflect a combination of amazement as well as bewilderment: "It's kind of hard to believe. It looks to me like something landed here . . . but if it landed here, I don't see any tracks . . . I don't know how anything could have landed and not leave tracks." These remarks alone rule out the Brazel debris field where numerous firsthand witnesses reported seeing a long gouge in the rocky terrain.

Noticing some of the remaining metallic-looking, foil-like wreckage scattered about, Rickett assumed that it was metal. "Is it hot (radioactive)? Can you touch it?" he asked Cavitt, who responded, "Yeah, be my guest, that's what I wanted you to ask me."[13]

Rickett then proceeded to pick up one of the pieces nearest him, which was about 4 inches by 10 inches in size. He described it as being extremely thin, featherlight in weight, and slightly curved in shape. After a cursory examination, he tried putting it over his knee in an attempt to make it bend. It would not, even in the slightest. "Cav looked over at [Easley] and said 'Smart guy . . . trying to do what we couldn't.'" Dumbfounded by this turn of events, Rickett exclaimed to Cavitt, "For God's sake! What in the hell is this stuff made of? It can't be plastic . . . don't feel like plastic . . . but it just flat feels like metal . . . I never saw a piece of metal that thin that you couldn't bend." Walking the site with Cavitt and Easley, he encountered more of the foil-like, but super-strong, metallic artifacts remarking, " . . . the more I looked at it, I couldn't imagine what it was."[14]

Indicative of the secrecy that surrounded the entire operation was a brief conversation that Rickett had with one of the MPs he recognized who was posted on the perimeter. Out of earshot of the others, he confided to Rickett, "I don't know what we're doing, but I do know this, I never talked to you in my life, not out here." Rickett's reply to this was, "What you see out here, you never saw it." "That's right," responded the MP. This sentiment was reemphasized to Rickett shortly thereafter by Cavitt in the presence of Major Easley in a manner reminiscent of a schoolteacher lecturing a remedial student. "You and I never saw this. You and I have never been out here. We don't see any military people out here. We don't see any vehicles out here." Rickett's reaction to all of this was, "Yeah, that's right. We never left the office . . . "[15]

Cavitt had just declared the incident as a non-event. It never happened. Such demonstrated the training of an intelligence officer.

Almost fifty years later, Cavitt allowed to Colonel Weaver the possibility that this conversation might have indeed taken place, but with a different, if convoluted intent. "Now I could have said something facetious like that after we got back to the office, after I was convinced that it was a weather balloon, or some such contraption . . . I could have said, 'Rickett, this has been a boon-doggle. I don't want 700 CIC Headquarters to know we wasted our time on it. Forget we ever did it.' I mean I could have said it in a facetious way: 'Let's make out like it never existed, because we're wasting our time.' But I didn't say it in such a way that it would be so highly classified we won't have anything to do with it."[16] Now, if anyone can make out what exactly Cavitt was trying to explain, please contact us. Cavitt, always the intelligence officer, was indeed well trained to spin his way out of any situation. Moreover, fortunately for him, Weaver knew nothing of the truth to refute him. He was assigned to find testimony to build a conventional case against Roswell—not educate himself to the contrary. And Cavitt was the perfect candidate—albeit the only one who fit their preconceived agenda.

As an example of Cavitt's contradictory testimony, consider what took place the very morning after Major Marcel returned from Fort Worth, Texas, where he had been ordered to pose for pictures with the substituted weather balloon in General Ramey's office. Marcel immediately confronted Cavitt about what had transpired during his two-day absence. Why would Marcel continue to force the issue after supposedly being proved wrong in his assess-ment of the wreckage while in Fort Worth? According to Rickett, Marcel asked the other officer for a look at his report on the incident. "Cav told him that he [Marcel] wasn't cleared to see it and couldn't look at it. Jess got upset at this and reminded Cavitt that he outranked him." Once again demonstrat-ing control, even over a superior officer, Cavitt told Marcel, as described by Rickett, "Cav told Marcel that his orders came from Washington and that Jess could take it up with the Pentagon. Marcel, who was a very diplomatic sort, dropped the matter and, as far as I know, never brought it up again."[17] Mind you, this dispute took place two days after Roswell had been explained away as a weather balloon.

In 1982, Cavitt acknowledged to investigative author William L. Moore that he probably had taken charge of things while Marcel was away in Fort Worth and that if such a report was written, he would have been the one that wrote it. "If there ever was a final report [on this matter], then I suppose I was the one who wrote it." But by the time Colonel Weaver interviewed him in 1994 that possibility had turned into denial: "I don't think I even made a report . . . which I normally would if there was anything at all unusual."[18]

Later in the afternoon, on the day of the actual incident, Cavitt, still in control, told Rickett to accompany him to the flight line to meet a plane that was arriving from Washington, DC. "They're coming in for that box,"[19] was all that Cavitt would say. This was in reference to a box of genuine wreckage, which Marcel had gathered, then taped shut before giving to Cavitt. According to Rickett, " . . . Cavitt gave him a receipt for it."[20]

Apparently, Cavitt wanted Rickett to witness the transaction. It would appear that the chain of custody needed to be witnessed. The plane, a four-engine C-54 as identified by Rickett, came in right as scheduled. Upon approaching the landed aircraft, Rickett would recognize the copilot, but for some unstated yet understood reason, neither spoke to the other on this occasion; secrecy was still the modus operandi. Cavitt personally handed over the box to two passengers from the plane that Rickett described as "CIC-types,"

CIC Captain Joseph T. Wirth in a late 1940s photo. (Photo courtesy of the Army Air Forces.)

who signed for it and boarded with the box. The destination: Andrews Air Field in Washington.

It would be later that year when Rickett was talking on the phone with Joseph T. Wirth of CIC Headquarters in Washington about an unrelated topic that he started to question Wirth about the previous July plane flight (he was the copilot). Before he could utter another syllable, Wirth cut him off. "Now you stop right there. No airplane came in there. Nobody got anything."[21]

It would take another fifteen years before Rickett had another opportunity.

During that time he had gone on to serve in the CIA/OSS and retired as a lieutenant colonel. At that time, Wirth was a civilian working for the Washington, DC, Park Police. Still remembering the abrupt last phone conversation back in 1947, the officer confided to Rickett this time, "You almost got me in a little bit of a bind, but I think we covered ourselves." Rickett agreed, " . . . somebody could get court-martialed for a thing like that."[22] Still wary fifteen years after the fact, Wirth suggested that the two of them go out into the parking lot to avoid the chance of being overheard. Though a civilian, Wirth still held a reserve commission.

Continuing the conversation, Wirth told Rickett, "Honest to God, they still haven't found out yet just what that [foil-like debris] was."[23] They then discussed the strange properties and possibilities of Monel Metal, which is similar in appearance to aluminum but much heavier, and concluded that beyond appearance, there were no other comparisons. Rickett then mentioned to Wirth that he had also run into an old Washington CIC colleague who later, like Cavitt, had transferred to the Air Force OSI. Rickett referred to him as only "Miller" during one of our interviews, but we determined that this was none other than the CIC agent who was responsible for the disbursement of funds for the recovery operation at the Roswell sites as a result of the crash in 1947. After exchanging pleasantries, mostly about the luck of discovering one another still alive, Claire Miller unexpectedly blurted out to a startled Lewis Rickett, "The answer's still the same. Don't ask me!"[24]

Although Rickett told us many times that he was sure that he didn't know the entire Roswell story, he did know enough firsthand information to conclude: "The Air Force's explanation that it was a balloon was totally untrue. It was not a balloon. I never did know for sure exactly what its purpose was, but it wasn't ours."[25]

Early into our investigation, we had submitted the names and documented serial numbers of high-level personnel at the RAAF in 1947, to both the DOD and the Veteran's Administration for further confirmation of service and in an effort to locate them. One of the names that could not be documented as serving in Roswell at the time in question was Army intelligence officer Ernest O. Powell. Former CIC/OSI agent Charles R. Shaw served under Major Ernest O. Powell at Selfridge Air Force Radar Station in

Mount Clemens, Michigan, in 1950. In two separate letters written by Shaw to former CIA employee Karl T. Pflock he stated:

> . . . [Powell] informed me he had been apprised by the G-2 [Assistant Chief of Staff, Intelligence] officer or the base provost marshal to the effect that an official source confirmed the recovery of alien bodies at the said UFO crash site. The most peculiar aspect of his [Powell's] assignment as OSI Detachment Commander at Selfridge AFB was its brevity, [commanding officers] including myself were puzzled by Powell's swift departure, without even a customary farewell party . . . [O]ur district [25] Commander, Colonel Raymond S. Rife [deceased] . . . advised that Powell was involved in a sensitive matter at his previous duty assignment [Roswell?], which necessitated his reassignment out of OSI. I never suspected Powell was previously stationed at Roswell Army Air Field. Therefore, the thought strikes me that his loose talk concerning the Roswell incident might have precipitated his transfer out of OSI.[26]

During his rehearsed interview with Colonel Weaver in 1994, Sheridan Cavitt made much of his claim that no one had ever contacted him or had sworn him to secrecy in any way regarding the events of 1947—contrary to telling Marcel upon his return to Roswell at the time of the event to " . . . take it up with Washington." Whether his claim is true or not is irrelevant. Historically, it was understood by CIC agents from the beginning of their highly specialized training that the life they would lead would be shrouded in secrecy and undercover work. It was precisely the reason they all dressed in civilian clothes and not in any style of uniform. In addition, although a select few officers at the RAAF knew their respective military ranks, most personnel had no idea. In other words, they lived a life of total secrecy and Cavitt was a master spy. He had seventeen years of high-level intelligence under his belt and knew how to keep secrets with the best of them. The very fact that Base Commander Blanchard originally sent him out with Marcel to accompany Brazel to the debris field is totally indicative of the high level of importance of the crash. He also knew that Cavitt could keep his mouth shut.

Contrary to what he later told Weaver, when he was first interviewed by researcher William Moore in 1982, Cavitt insisted that he was "probably still bound by his security oath and the matter was probably still classified or at least it certainly was at the time."[27] They did not have to be told not to talk in different situations; it was clearly understood from day one not to talk about classified topics—even to spouses and families. His late wife Mary, contrary to her own statements to Weaver about ignorance, perhaps summed it up best during an interview with Moore when her husband briefly stepped out of the room. "He won't tell you anything. They've told him not to, and he won't. That's why they chose him for many of the assignments he had . . . because he doesn't tell anything."[28] It remains curious that after our first visit and interview of the Cavitts at their then winter home in Sierra Vista, Arizona, in 1990, Cavitt walked us out to our car. After an entire afternoon of denial and obfuscation, we were beyond frustrated and could foresee that the stubborn former CIC officer would be a tough nut to crack. But what transpired next spoke volumes to us. "Boys," Cavitt addressed us as we stood outside the car, "I just need to know one thing: if you should find anything that threatens national security, you will keep it to yourselves." For a moment we glanced between ourselves and looking back at Cavitt we then responded with, "Yes, Colonel. Of course we will, Colonel." Cavitt seemed pleased as he replied, "Good. I'm happy to hear that." No sooner had we exited down the driveway than we again looked at one another and said, "He knows." On numerous occasions thereafter we began to hear about attempts within Cavitt's own family to get him to admit anything about Roswell. In such cases he would either storm away from the Thanksgiving table or simply pretend he didn't hear or understand the question. Such gamesmanship played out right up until his death as described in Chapter 23.

Given Sheridan Cavitt's lifetime of denial and altering the facts concerning his participation in the Roswell Incident, and given the caliber and volume of credible testimony in direct conflict with his own contemporaries including his own family, he must be rejected as a credible witness. This would apply to both sides of the coin, including those who still subscribe to the Project Mogul explanation. Sheridan Cavitt is presently the only witness to their folly and he has been summarily disqualified.

Shortly before his death, Lewis Rickett received an unexpected phone call from his former boss with whom he had not spoken since 1950. Cavitt announced that he had called to wish him a happy birthday after forty years. The conversation quickly switched to 1947 and Cavitt cautioned his former subordinate who, by this time, had become a "cooperating witness'" for Roswell researchers, "Who have you been talking to?" Rickett answered, "Don Schmitt," to which Cavitt remarked, "Oh, yes, I know Don. We've talked as well. We both know what happened out there." Rickett's animated response was, "We sure do! What do you plan to tell them?" Cavitt said, "Well . . . maybe someday. Goodbye, Bill!" Those were the last words Rickett ever heard from the one person who shared so many of the events from 1947 back in Roswell. Cavitt abruptly hung up the phone.[29]

Just before passing away in 2007, Colonel Doyle "Dode" Rees was found by his daughter Julie sitting in a chair at a window at her home in Utah just "staring out at the sky." Curious, she asked him, "What are you looking for, Daddy?" His answer was simple: "I'm looking for UFOs. They're real, you know." Julie, taken aback by this spontaneous revelation, wasn't prepared for her father's next remark: "I saw the bodies." With that, he rose up from the chair and left the room, never to reveal another detail.[30]

Then Lt. Colonel, Doyle "Dode" Rees was Cavitt's boss in 1947. Stationed at USA/OSI at Kirkland AFB in Albuquerque, New Mexico, at the time of the Roswell Incident, he kindly wrote a letter at our request to Cavitt in 1991. In it, Rees remarked, "When you call the press conference to tell the world [about Roswell], let me know, because I want to be there."[31]

Lt. Colonel Sheridan W. Cavitt died on December 30, 1998, at the age of seventy-nine. A faithful soldier to the very end, he took all of the truth with him. There was no press conference.

3

THE CORONA DEBRIS FIELD: MUCH ADO ABOUT SOMETHING

Based on all supporting information regarding the Mogul balloon explanation, this must have been an extremely isolated occurrence. After all, how could something as mundane as a weather balloon have generated all the excitement it evidently did? And indeed, if this happened to be a specific Mogul launch gone awry, which project directors would later categorize as *missing*, how was it that so many people would become involved in the immediate ranchland surrounding the debris field location? How was it that so many civilians were able to find remnants of a top secret project while the authorities, who now claim they *were* looking for it, could not? And how is it that none of them describe the remains of any type of balloon device, top secret or otherwise?

Anyone who has ever taken the time to travel to this remote territory immediately observes the immense size of the ranches themselves. One would also learn that, in 1947, this area of New Mexico had little if any electricity—in other words, no TV or radio: a total lack of outside communication. It is fascinating to note that this particular region didn't even receive phone service until 1986! This is harsh, high-desert country by any definition—brutally hot in the summer and frigidly cold each winter. Prevailing winds are cause for constant soil erosion, and much of the land is barren and stark. Rattlesnakes, tarantulas, scorpions, and Gila monsters all share a space with the local human population. Monsoon rains typically arrive each July and August, causing gravel and dirt roads to wash out and become impassible. Even today, four-wheel-drive vehicles are required transportation through much of the region. Broken axles, drive shafts, and flat tires are the norm. The land has basically remained the same throughout the centuries and is famous for the Lincoln County range wars, which followed the War between the States. Billy the Kid, Pat Garrett, and John Chisum rode the same territory that Mack Brazel and all his neighbors did in 1947. It continues to amaze us how the most ardent of skeptics, regardless of never having spoken to a single witness, think that residing in or passing through Albuquerque fully qualifies them to make an educated opinion on ranching the high country of central New Mexico.

That is why we intend to demonstrate to the reader all of the ancillary activity surrounding this incident in the Corona ranch region, which *did not involve Mack Brazel or his family*. Why would so many adults and young people travel in pickup trucks or ride on horseback up to seventy-five miles on dirt roads to witness a downed weather balloon every one of them had stumbled across so many times before? Even today, there is an old water-holding tank that remains as direct testimony to many decades past—it is overflowing with rotting weather balloons gathered up by young and old alike from one end of the ranch to the other. Any responsible rancher will tell you that there is absolutely no way that Mack Brazel would allow the remains of any type of discarded wreckage, Mogul included, to be left lying out in that open pasture. You see, cattle and sheep are like goats. They'll eat anything in their path, and all that neoprene rubber from cluster balloons in a Mogul device could suffocate the poor animals. Keeping the herds alive and productive was the primary function of the ranch supervisor, and there is no reason to doubt that Brazel was good at his job. But this situation was much different. There was so much of the stuff, and nowhere was there an instrument package or even a name tag to be found directing the bearer to whom and where to report the find (as would be expected with weather research balloons, particularly the Mogul balloons, which had their own "Reward Notice" from New York University). And besides, the "damn stuff" scared the sheep away from the pasture anyway.[1]

Also keep in mind that this was before the military was ever alerted to the find, and days before the Army would issue the now-famous press release about the capture of a flying saucer. In fact, it was *three or four days* earlier.

So who are these other witnesses the Air Force now refuses to acknowledge? Who are these other people who were drawn to see a whole pasture full of super-strong material that snapped back after being bent? And what happened to them as a result of seeing too much?

The very first outside witness just happened to be with Mack Brazel the very morning he discovered the wreckage. We have confirmed from numerous sources that this young boy would often spend weekends with Mack learning to be a good ranch hand. He was the seven-year-old son of Mack's nearest neighbors, Floyd and Loretta Proctor. His name was Timothy, but his family and friends called him "Dee" after his middle initial. After Mack returned him home with a sampling of the wreckage, as soon as he had the chance, he got right back on his horse and gathered up a number of friends and headed back to explore the wild story all the adults were talking about. There appears to have been a young girl who accompanied the "cowboys." A few have mentioned her by name, but she still refuses to talk to this day.

Such would also become the fate of Dee. Whatever else he witnessed, according to his mother, Loretta, in days to come, he would return home and behave as though something or somebody had severely frightened him. And from that moment on, nobody, including his family, could get him to utter another word about the adventure—an adventure that, by all accounts, tormented Dee for the rest of his life. True, many years later he would dismiss the entire affair and chuckle that he wasn't even born at that time—which got quite a reaction from his mother. But he would finally share a glimpse of the true story with her in 1994 when she was suffering from a life-threatening blood clot in her neck. Dee decided to pick up his mother and take her for a drive—not to the site everyone already knew about, but rather atop a ridge some two and a half miles to the east. Even then he parsed his words carefully, but he wanted her to know just in case. "Here is where Mack found something else," he confessed. His mother couldn't get her own son to admit anything further. And he never did. Dee Proctor died of a heart attack in January of 2006.[2]

One of the other young men who we believe was with the others was the son of a local hired hand. He went by the name of Jack but his correct name was Sydney Wright. It was in 1998 that Jack would finally disclose what may also have been the cause of Dee Proctor's lifetime of silence. Jack admitted that he, the two sons of rancher Thomas Edington, and one of rancher Truman Pierce's daughters somehow managed to get to "the other location."[3] It was there that they witnessed something that none of them expected.[4] Later, this would most certainly play into the military abduction of Mack Brazel.

There can be no doubt that Brazel made every effort to enlist the advice of many of his neighbors. Ranchers such as Clint Sultemeier, curious enough about the story, would drive over to the Foster ranch (of which Brazel was the custodian) and actually retrieve a number of souvenirs. Sultemeier mistakenly thought he was smart by hiding his collection where nobody would think of looking. But officials would soon be retrieving *all* the evidence, and let nothing stand in their way in doing so.[5]

Within the next two days after the crash, others who owned surrounding ranches would go out of their way to check out the story about "pieces of a flying saucer." Budd Eppers and Truman Pierce would arrive on the scene.[6] Glaze Sacra would load a number of "weightless" pieces of metal into his pickup and head discreetly home.[7] Danny Boswell's parents, who owned a ranch twenty-five miles to the east, drove forty-five minutes to see for themselves what everyone was talking about.[8]

Just what everyone was talking about were not pieces of a Mogul weather balloon. Rather, eyewitness descriptions were consistent about nearly indestructible characteristics, fiber optics, I-beams, and the pieces that amazed them beyond all others—the "stuff that would flow like water after you wadded it up in your hands."[9] In any event, the evidence remained scattered across the open range of the Foster ranch as prevailing winds continued to whisk it more and more to the southeast for *three full days* before Mack Brazel would finally complete his weekly chores and make the seventy-five-mile trek to Roswell and report the fantastic discovery to the authorities. But not before he had exhausted every attempt to get every last word of advice he could. Besides the Proctors and Sultemeiers, he took samples of the debris to his uncle Hollis Wilson, and perplexed patrons at Wade's Bar in Corona would hand fragments from one end of the bar to the other, each one attempting to cut or burn the pieces.

From there he would show the owners as well as customers at the Corona General Store. It seemed as though no one had any solution to what crashed out on that arroyo. From all known accounts, it would seem that everyone remained as puzzled as the next person. Portions of the unknown object eventually found their way to the annual Fourth of July rodeo an hour to the south in Capitan. Holiday revelers recalled memory metal being flashed around the festivities. It seemed as though everyone was aware of the crash, *except the authorities!*

Finally, the frustrated Brazel would run into his friend, State Police Officer Robert Scroggins. At last, someone with an official background who might be able to shed some light on the entire mystery. But no, the experienced officer had never seen anything like it either. The best he could offer was an alternative plan of action: He was heading home to Hobbs, which was east of Roswell. He would take a larger section of the memory material and report it to the military in Roswell. Yet, Scroggins couldn't help but wonder, "If it was military, why weren't they looking for it? And if it wasn't ours, then whose was it?"[10]

Still, Brazel remained undeterred and became more and more agitated as to under whose purview the wreckage remained. Someone was responsible for nearly a mile-long swath of the stuff, and he couldn't get the sheep to cross through the area to get to water. And so he was left with no alternative but to make the long drive down to Roswell and report it himself. After all, maybe some of his neighbors might be right about the reward for a genuine "flying saucer," and he did have ranch business to attend to anyway. So, Brazel made the tedious journey to Roswell first thing Sunday morning, July 6. Little did his friends and neighbors suspect that Brazel would be getting into the territory of military regulations and restrictions that would alter his and his family's perception of patriotic duty. Brazel was about to experience a dark side of his country he had never imagined.[11]

No sooner had word hit the community of Roswell about all the "strange goings-on" up at "some ranch north of town" than others saw an opportunity for possible fame and fortune. Charlie Schmid, who lived just north of town, jumped right on his motorcycle and headed out into the desert. "I had no idea what to expect, but it sounded like the thing to do at the time," mused Schmid. "I somehow managed to skirt around enough fence lines, sinkholes, and livestock, and arrived at the outer rim of all that metal. I picked up a few pieces along the fence. They had funny writing on them and seemed real strong. I thought it had to be some big secret test by the army, so when I heard an engine approaching I just took off. It was definitely no weather balloon."[12]

Paul Price and his older brother also heard about the crash of a "flying saucer" north of town. "We knew most of the ranchers, so it wasn't long before we got to the right spot. There were so many parts for as far as you could see. Some of the pieces just snapped back in your hands when you bent them." Price was asked if he and his brother retrieved any of the pieces. "I have nothing more to say about it," he replied.[13]

At sunrise on Monday, July 7, Major Jesse Marcel and Captain Sheridan Cavitt, two intelligence officers from the RAAF, would arrive on the scene. Brazel and the two men had just spent the night at the old Hines house after the rancher brought them there the evening before. They would spend the better part of the day examining, collecting, loading, and containing what they couldn't fit into two vehicles. Marcel would send Cavitt on ahead to report directly to Colonel Blanchard back at the base. The major would check over the area one last time and head out later. All testimony from firsthand witnesses maintains that debris still covered an area the size of a football field by the end of the day on Monday. More troops would be needed. A full-scale cleanup operation would then commence with fifty to sixty troops the very next morning.[14]

One of the young sons of a hired hand from the Richards ranch, which adjoined the Foster ranch from the south, spied with a couple of other boys from a distant hill. His name was Trinidad "Trini" Chavez, and he said, "There were soldiers lined up and picking up all this material. Trucks and jeeps surrounded the area. We saw men with rifles get out of one of the trucks; well, we figured we saw enough." Young Dan Richards had already recovered a few of the memory material pieces, and unknown to Trini at the time, his dad had as well. Alas, it was too late for Trini to take a piece of the wreckage himself. "Too many damn soldiers," Trini said.[15] Later that day, after dark, witnesses would report observing trucks with large spotlights heading in earnest to the "other location."[16]

It would appear that the military extended its cordon of the area to not only the immediate debris field, but also the ranch proper, the main ranch house, and the outlying roads. Budd Payne, another rancher in the Corona area, found out the hard way by merely chasing a stray steer onto the Foster ranch: no sooner had he ridden onto the property than a jeep carrying MPs roared over a ridgeline and bore down on him. Payne, who would later become a county judge, was physically accosted and forced off

The "Hines House" on the J. B. Foster Ranch is now gone. (Photo courtesy of Tom Carey.)

the ranch. He was ordered not to set one foot on that parcel of land until their work was completed. Years later, at Payne's constant encouragement, Brazel would show him the precise location "which caused all that commotion."[17]

Now, the military was fully aware of all the civilian curiosity. They had to account for each piece, and they suspected everyone—*including children!*— of taking debris. Ranchers were forced to inform on one another. Ranch houses were searched and ransacked. The wooden floors of livestock sheds were pried lose plank by plank and underground cold storage fruit cellars were emptied of all their contents. Glass jars were scattered, broken on the ground.[18]

Corona ranch hand Trini Chavez, seen here in a 2005 photo, was a teenager in 1947 who clandestinely witnessed the military cleanup of the debris field site on the Foster Ranch with his friend Dan Richards. (Photo courtesy of Tom Carey.)

American citizens were threatened into submission by agents of the U.S. government. Not a single shred of evidence could be left behind, though some would later claim that a piece of the memory material was kept in the safe at the pumping station a mile away.[19] Troops spent more than two days picking up every last trace in the pasture, finishing with industrial vacuum cleaners.[20] All that would remain were a gouge in the land and the tire tracks that crisscrossed the rocky terrain. Just as the ground had been stripped naked of all its secrets, so had the local residents.

A vast assortment of physical proof was recovered from more civilians than we will ever know. Some may have temporarily slipped through the military's grasp, but though we have had numerous false alarms, no such evidence has surfaced and no one is talking. Stories about home and vehicle break-ins abound even up to the 1990s, and so do the mysterious disappearances of actual debris. Rancher L. D. Sparks described how, just a few years after the original incident, Dan Richards had him toss a thin piece of foil-like material in the air as he fired a rifle at it. "Shot after shot just ricocheted off of it," Sparks claimed. "I would crumble the piece into a ball and watch in amazement as it would unfold as it floated through the air."[21]

Unfortunately, we'll never know what became of his proof. Richards was killed in a single-vehicle accident shortly thereafter. And, true, it is very conceivable that the long chain of caves that surrounds his parents' ranch house may have been too large a haystack even for the military—as it remains for present researchers. Deep in the bowels of the old Richards ranch, buried treasure may still lie yet to be discovered.

Everything was soon *forced* to return to normal. "You are not to say another word about the incident" became an all too common catchphrase throughout the Corona high-desert region. It seems unlikely that an event that captivated the nation, affecting the personal lives of so many and leaving them cold and cynical about their government, could be a simple weather balloon device. What tactics were utilized to ensure the full cooperation of all the civilians who saw and knew too much? Roswell photographer Jack Rodden did business with many of the ranchers from the Corona area. He says that one of the ranchers told him that his three kids had come home one day, at the time of the incident, frightened to death and refusing to talk about what they had seen. Rodden pressed the issue and was told by the old-timer that the kids had gotten too close to something, and someone in the military had scared them badly.[22] Other parents emotionally recounted how their young children were never the same. "When they returned home they looked as though they had seen a ghost!" "They were frightened, shocked, and grew increasingly paranoid," spoke a number of the residents. About what? A Mogul weather balloon? Observing a field peppered with strange wreckage? Or was it what a handful of them regrettably saw at the "other location"? Was the military forced to take them aside and put some additional fear of God into them? Almost all of them to this day refuse to talk about it—even with their families.[23]

Many years later, Sydney "Jack" Wright would lament how he grew up overnight from the image that has haunted him all these years: "There were bodies, small bodies with big heads and eyes. And Mack was there too. We couldn't get away from there fast enough."[24]

4

"THEY'RE NOT HUMAN!"

From the very beginning of the Roswell investigation by civilian researchers more than two decades ago and until quite recently, the role played in
the Roswell Incident by the sheep rancher who started it all, Mack Brazel,
had pretty much remained unchanged. As will be covered in detail later in
the book, sometime during the first week of July 1947, in the course of his
duties as the foreman of the J. B. Foster ranch, located 33.2 miles southeast
of Corona, he discovered a pasture full of pieces of strange, silvery wreckage
the morning after a severe thunder and lightning storm during which he
thought he had heard an explosion. After consulting with neighbors and relatives about his find and being told about a possible $3,000 reward offered by
a newspaper for pieces of a "flying saucer,"[1] he took their advice and drove the
seventy-five miles in his ancient pickup to Roswell.[2]

Upon his arrival in Roswell, Brazel reported his find to the sheriff's office.
Deputy Sheriff B. A. "Bernie" Clark had the weekend duty and received Brazel.
Puzzled by what Brazel was showing him, Clark thought about it and finally
bit the bullet and called for the sheriff, George M. Wilcox, who lived with his
family in the upstairs residence over the jailhouse. Not happy about having his
Sunday interrupted, Sheriff Wilcox appeared anything but interested enough to
do anything about it. Brazel and Clark showed him some samples of the wreckage Brazel had brought with him, but to no avail. Fortunately for Wilcox, whose
eyes had glazed over, he then received a phone call from a young announcer
from Roswell radio station KGFL named Frank Joyce, who was looking for any
local tidbits of news that he might put on the air—not a minute too soon for a
thankful Wilcox as he passed Brazel off to Joyce with, "Well, there's somebody
here right now with a story you might be interested in." So Wilcox handed

KGFL Roswell radio announcer and United Press "stringer" Frank Joyce at the mic in a 1947 photo. (Photo courtesy of Frank Joyce.)

Brazel the phone. After some back and forth between the two, Joyce suggested to Brazel that he should contact the air base in town, and hung up—or so we thought.

We knew that two days later, on July 8, 1947, the *Roswell Daily Record* carried a banner headline heard 'round the world: "RAAF Captures Flying Saucer on Ranch in Roswell Region," which was retracted the next day, when the Army Air Forces claimed that the flying saucer wreckage was really just the misidentified remains of a weather balloon. We also knew from family members and friends that during this period, Mack Brazel had been taken into custody and detained at the RAAF for the better part of a week or more. He also seemed to have changed his story from that of the previous day, as evidenced by a front-page interview article that appeared in the July 9, 1947, *Roswell Daily Record*, wherein he described his find in terms consistent with a rubber weather balloon and a tinfoil radar target. At that point, the press lost interest in the story, and Brazel was allowed to go home a few days later, a bitter man at having been treated so badly by his country just for having done what he saw as his patriotic duty.

There were several things about Brazel's account that always bothered us. One was the persistent claim by some of his neighbors that, shortly after he was released from the clutches of the military, he was somehow able to purchase a brand-new pickup truck and then leave his Foster ranch employment entirely to start his own business in Alamogordo, New Mexico, which was closer to his home in Tularosa. Not bad for someone barely scratching out a living off the land and known for not having two nickels to rub together. In short, it had all the earmarks of a bribe to us. Another mystery was the fact that the military took Brazel into custody for a week or more and basically worked him over. Upon his release, Brazel bitterly told his family that he felt as if he had been in jail and that he would never again report anything to our government "unless it was an atomic bomb." We could not understand why the military would go to such lengths with Brazel, as it had done with no other witness, if he had only discovered pieces of wreckage—no matter how strange. Surely he could have been convinced in short order that it was from a secret project or some such cover story, and Brazel could have gone on his way a happy camper. Something was missing, it seemed to us, in Brazel's story.

The first hint at an answer to our questions was provided in a series of Roswell update articles by William Moore in the early and mid-1980s, before Moore left the Roswell case. He had located and interviewed KGFL announcer Frank Joyce. By that time, Joyce had left Roswell and had become a well-known radio and TV personality at KOB in Albuquerque. Joyce told Moore that he had had not one, but *four* conversations with Brazel. We knew about the first one between Brazel and Joyce on the telephone, but the other contacts were new information.

Frank Joyce seen in a 1998 photo when he told the authors "the rest of the story" about his interviews with Mack Brazel. (Photo courtesy of Tom Carey.)

In May of 1998, we paid a visit to Frank Joyce in his Albuquerque home. He had recently retired from KOB and from public life and seemed to be in a talkative mood. He recalled for us that he had never spoken to anyone about his 1947 Roswell experiences for the first twenty years after the event. When he finally did start to talk about it, he always stated that there was more to his story, but that he did not want to talk about it at the time. Now, post-retirement, he was feeling differently about things. He told us that he was going to tell us something he had never told anyone, and started us off on a cautionary note: "Don't stop me once I get started, or I might realize what I am doing and shut up." We were only too happy to oblige. The following is our reconstruction of Frank Joyce's account of his initial conversation with Mack Brazel on the afternoon of July 6, 1947, shortly after Brazel had arrived in Roswell to report his discovery, as revealed to us for the first time by Frank Joyce.[3]

BRAZEL: [angrily] Who's gonna clean all that stuff up? That's what I wanna know. I need someone out there to clean it up.

JOYCE: What stuff? What are you talking about?

BRAZEL: [somberly] Don't know. Don't know what it is. Maybe it's from one of them "flying saucer" things.

JOYCE: Oh, really? Then you should call the air base. They are responsible for everything that flies in the air. They should be able to help you or tell you what it is.

BRAZEL: [At this point, according to Joyce, Brazel really started "losing it."] Oh, God, oh, my God. What am I gonna do? It's horrible. Horrible. Just horrible.

JOYCE: What's that? What's horrible? What are you talking about?

BRAZEL: The stench. Just awful.

JOYCE: Stench? From what? What are you talking about?

BRAZEL: They're dead.

JOYCE: What? Who's dead?

BRAZEL: Little people. [now barely audible] Unfortunate little creatures. . . .

JOYCE: [At this point, Joyce thought to himself, "This is crazy!" He decided to play the role of devil's advocate to a story he did not believe.] What the . . . ? Where? Where did you find them?

BRAZEL: Someplace else.

JOYCE: Well, you know, the military is always firing rockets and experimenting with monkeys and things. So, maybe . . .

BRAZEL: [shouting now] God dammit! They're not monkeys, and THEY'RE NOT HUMAN!! [With that, Brazel angrily slammed down the phone to end the conversation.]

KGFL had every intention of broadcasting the story of the century, which is why its staff escorted Brazel to the home of the station's owner, Walt Whitmore Sr., and recorded Brazel's testimony late in the evening of Monday, July 7, 1947. At least, that was the plan—until the U.S. Army took custody of Brazel and the KGFL wire recording, and removed both to Roswell Army Air Field south of town.

Efforts continued the next morning to disseminate preliminary news information to the local townspeople. But Washington was watching. Early in the morning on July 8, George "Jud" Roberts, minority owner at the station, received a long-distance phone call from T. J. Slowie, the executive secretary of the FCC, who warned him that the matter involved national security. Should KGFL air any portion of Brazel's interview or issue any information regarding it, it would lose its broadcasting license.[4]

As if that weren't enough to squelch the story, another call to KGFL came from Washington a few minutes later. It was from U.S. Senator Dennis Chavez, who strongly suggested that KGFL do exactly as the FCC had cautioned.[5] When station executives asked for his help, he indicated that the decision was out of his hands. The station immediately complied with the FCC's order.[6]

While the officially sanctioned intimidation of a news source in Roswell was underway, another situation was developing at KOAT Radio in Albuquerque, an affiliate of both ABC and Mutual networks at that time. Secretary Lydia Sleppy remembered vividly the frantic phone call she received from John McBoyle, general manager and part owner of sister station KSWS in Roswell, which had to rely on KOAT to transmit to the Associated Press wire service. "Lydia, get ready for a scoop!" McBoyle excitedly said. "We want to get this on the wire right away. Listen to this! A *flying saucer* has crashed. . . . No, I'm not joking. It crashed near Roswell!"

Sleppy urgently asked program director and acting station manager Karl Lambertz to witness her reception of the story and its transmission. Using the teletype, Sleppy alerted ABC News headquarters in Hollywood to expect a "high bulletin" story. Lambertz looked on as she initiated the connection. "It's a big crumpled dishpan . . . . ," McBoyle—hardly containing himself—continued over the phone, ". . . and get this. They're saying something about little men being onboard."

Before Sleppy could type out a mere couple of sentences, a bell rang on the teletype machine, indicating an outside interruption. McBoyle, meanwhile, started to converse with someone in the background and the discussion became more intense as it went along. Moments later he nervously told Sleppy, "Wait a minute, I'll get back to you . . . . Wait . . . . I'll get right back."[7]

He did not. The very next moment, the teletype came back on line and printed out the following order:

ATTENTION ALBUQUERQUE: DO NOT TRANSMIT, REPEAT, DO NOT TRANSMIT THIS MESSAGE. STOP COMMUNICATION IMMEDIATELY. NATIONAL SECURITY MATTER.

In stunned disbelief, Sleppy observed that the message was from the FBI. No further attempt was made to transmit McBoyle's amazing story in any shape or form.[8]

At a later time, Sleppy was speaking with McBoyle and broached the subject of the strange series of events. The veteran reporter's response shocked her: "Forget about it. You never heard it. Look, you're not supposed to know. Don't talk about it to anyone." Another time, he mentioned to Sleppy that he had observed a plane take off from the RAAF on route to Wright Field with wreckage on board, but was unable to get near it because of all the armed guards posted in the area.[9]

Lydia Sleppy would wait twenty-five years before breaking her silence. For his part, McBoyle returned to ranching in Idaho and, disregarding any and all exhortations from his wife, son, and daughter, refused to discuss the matter. "I don't remember," is all he would say to us before he died, thus taking the true story about the "crumpled dishpan" with him to his grave in 1991.[10]

Unfortunately, "they" soon got to Mack Brazel as well, and the next time Frank Joyce saw the rancher, he didn't have the same things to say. When the military escorted Brazel to KGFL, Brazel sat down at the microphone and retracted his original story. The unusual material that Brazel had carried with him in two boxes all the way from his ranch to Roswell was now "nothing more than a weather balloon," according to Brazel. Taking Brazel out of the broadcast booth during a music break, Joyce followed the older man out into the front lobby of the radio station. "That's not the story you told me before," Joyce bluntly told Brazel. The rancher stuck to his new story while growing more agitated. Brazel could see by the look on his face that Joyce was rapidly losing respect for him. Brazel then said, "They told me it would go hard on me if I didn't do what they said." Presumably, Brazel had been warned of the dire consequences—not to the nation, but to Brazel and his family—if he said anything that conflicted with the Army Air Force's new, official story. At that point, Joyce noticed the uniformed men standing just outside the glass door entrance. The reporter made one last attempt to get the truth. "What about the 'little green men' you told me about the other day?"

The rancher paused as he walked over to the door and put his hand on the doorknob. Turning toward Joyce, he casually said in a soft-spoken, matter-of-fact voice, "They weren't green," and out he went.[11]

5

"AFRAID THEY WOULD SHOOT AT US"

Robin Adair, a photographer with the Associated Press (AP), received a phone call from the main office in New York on Tuesday, July 8, 1947, telling him to get to Roswell, New Mexico, immediately, "even if it meant leasing a plane" for the journey from El Paso, Texas.[1]

Fully briefed on the situation just to the northwest, Adair felt it wise to try getting some aerial shots before landing in Roswell. He instructed his pilot to fly the plane north toward Lincoln County. He told us:

> We didn't do a bit of good by it. We couldn't get any [pictures]. Even then, the place was surrounded by policemen and FBI people. They wouldn't let us get within three quarters of a mile of the place. We were afraid they would shoot at us. We did take a plane up there, but we couldn't land anywhere around [the debris field on the Foster ranch]. We got as close as we could and we wanted to get lower [but the military officers on the ground] just waved. You couldn't tell if they were waving us off or just politely telling us to get the hell away from there.[2]

From the air, Adair managed to observe all of the activity at what was later determined to be the debris field. Many troops, vehicles, and MPs covered the large open field. Some areas also appeared to be scorched. Even from the altitude at which they were flying, the photographer could make out what he called the "gouge." He remarked, "You couldn't see too good from the air. Apparently, the way it cut into [the ground], whatever hit the ground wasn't wood or something soft. It looked like it was metal." His lasting impression was that it had descended, impacted the ground, and then ascended back into the air.[3]

A scout plane similar to the one that flew AP photo technician Robin Adair over the UFO crash site on his flight from El Paso, Texas, on July 8, 1947. (Photo courtesy of the U.S. Air Force.)

Heading south to Roswell, he observed that the terrain became more rugged and canyon-like. Still, he and the pilot continued to look for any type of military activity below. Adair said that he saw two recovery sites. "One of them wasn't very distinct. The other was [more easily seen]."

After landing at the old municipal airstrip west of Roswell, Adair linked up with reporter Jason Kellahin. Kellahin had also received a call from the New York office and had driven down from Albuquerque. That evening, the team went to the offices of the *Roswell Daily Record*, where Adair proceeded to set up the equipment to transmit wire messages back to Albuquerque.

The two planned to interview rancher Mack Brazel, who was now retracting his original story, though he continued to insist that what he saw was not any type of weather balloon.[4] Adair snapped the cowboy's picture—cowboy hat and all. The photo and Kellahin's story were tediously wired back to the New York office. Transmitting the pictures was considered so important, or maybe such a novelty, that the *Record* ran a story on the front page of the July 9 edition along with a photo of both Adair and Kellahin. The photo taken of Brazel during this session—a head shot of the wary Brazel wearing his cowboy hat pulled back slightly—was the first wire photo ever sent from Roswell by any news organization.[5]

Brazel's name and face would be household items, however briefly, by the end of the next day. Brazel told the journalists that he had found the debris almost a month earlier, on June 14, while doing his chores on the ranch. His wife, Margaret, daughter Bessie, and son Vernon were with him on the ranch at the time. In that interview (published in the July 9 edition of the *Roswell Daily Record*) Brazel claimed that the wreckage he found consisted of rubber, tinfoil, tape, and wooden sticks that were all confined to a rather small area. The story ended with Brazel saying that he had found weather balloons on two previous occasions, but that this object did not resemble any of those. "I am sure that what I found was not any weather observation balloon," he said.[6]

According to the *Record* editor at the time, Paul McEvoy, the military officers then escorted the rancher out of the news office immediately upon the conclusion of the interview. While they were walking toward the waiting staff car, two of Brazel's neighbors, Floyd Proctor and Lyman Strickland, passed by. Both men were surprised when their friend walked right past them without acknowledging them in any way. Proctor would later say that the military was keeping Brazel on a "short leash."[7] Three other neighbors—Leonard "Pete" Porter, who lived on the ranch just south of Brazel's, and ranchers Bill Jenkins and L. D. Sparks—reported that they saw Brazel "surrounded" by military personnel in downtown Roswell. They said that Brazel kept his eyes down and pretended he didn't notice them.

Unknown to Robin Adair and Jason Kellahin, who were chasing a big story that had come across the AP wire, they had competition from an unlikely source. Someone else was also coming for the story—armed and ready! Allan Grant, who passed away in 2008 at the age of eighty-eight, was an acclaimed photographer for a national magazine with many cover shots and awards to his credit. In the glory days of the large, photo-oriented, weekly "mags" such as *Life, Look, Collier's, McCall's*, and so on, photographers such as Allan Grant had extraordinary access to the glamorous worlds of fashion and Hollywood, as well as history-making events. In this capacity, Grant would do the last photo shoot of Marilyn Monroe before her death, get the first photographs of Marina Oswald only hours after her husband had assassinated President Kennedy, photograph the atomic bomb tests in the Nevada desert during the early 1950s, and photograph the only flight of Howard Hughes's infamous "Spruce Goose," a large float-plane made completely of wood, in 1947.[8]

In July of 1947, Allan Grant was a young staff photographer for *Life* magazine who had recently relocated to Los Angeles from New York to be near the Hollywood scene. According to Grant's account on a website that chronicles his life's work (no pun intended), on July 7, 1947, he was sent to New Mexico on what his editor termed a "meteor hunt." All he was told was that a "huge meteor" had crashed somewhere in the vicinity of Roswell, New Mexico, and that the Air Force was searching for it. He was assigned a pilot, who met him in Albuquerque, who would fly and land him in the vicinity of the crash site. To Grant's surprise, the pilot turned out to be a military pilot, Army Air Forces Major Charles Phillips. (Phillips would later work with Dr. Lincoln LaPaz as part of a "ground survey" team investigating the mysterious "green fireball" phenomenon that afflicted New Mexico a year or so after the Roswell event.) After a flight lasting thirty to forty minutes, Phillips put the single-engine trainer down on a "makeshift runway" located in the middle of the desert, most likely somewhere east-southeast of Albuquerque. (NOTE: There are a number of such small airstrips that dot the state of New Mexico, some still operating, and some remnants of a long-forgotten past. A flight from Albuquerque to Roswell today by a twin engine Beechcraft takes about fifty-five minutes.) Upon alighting from the aircraft, Phillips handed Grant a loaded semiautomatic pistol in a holster. The stunned Grant asked, "What do I need that for?" "My orders were to see that you were armed," replied Phillips. "Against what?" the now alarmed Grant asked, as he buckled the gun belt around his waist. Smiling, Phillips looked at Grant and just shrugged his shoulders, "We don't know."

The men got into a waiting jeep and drove off into the desert in a cloud of dust on a mysterious search for an unknown but potentially dangerous quarry. Years later, Grant reminisced about his strange trip: "Of course we never found the 'meteor,' and I have often wondered if I had arrived too early or too late. Or perhaps I was landed in the wrong place. At any rate, the story never ran in *Life*."[9]

In 2007, our associate Anthony Bragalia located Grant's widow, Karin, to try to glean additional information about her husband's July 1947 Roswell trip, and our investigation followed up with her with additional questions later. Mrs. Grant is a "*Life* alumna" herself, so she knows the business and can offer informed opinions as to what might have been taking place regarding her late husband's "flight to nowhere." According to Mrs. Grant, she met her future husband in 1966 when she arrived at the *Life* Beverly Hills office "fresh out of boot camp" in New York City. They were married in 1971. She said that her husband would be teased in the office each time there was a UFO sighting or something in the press about UFOs, and someone would invariably ask him, "You *did* keep that gun they gave you the last time, didn't you?" She recalls that it was in the 1966–67 time frame that her future husband first told her about his near brush with destiny in 1947, and she definitely remembers him saying the name "Roswell." According to her, the sequence of events involving her husband started when the *Life* magazine offices in New York City were alerted by an Air Force official in Washington, D.C., that a strange object, possibly a meteor, had crashed near Roswell, New Mexico. His orders were to see and shoot what was there. "When you get the call, you go wherever they say—on the double!" Except for the pilot, Allan Grant was alone on this trip. Mrs. Grant adds, "This would make sense if *Life* got the call from the Air Force. All they needed were some photos, and then a 'source' could fill them in on the text."[10]

So, was Allan Grant's July 7, 1947, "desert adventure" the result of just someone's misunderstanding or mistake, a military "snafu" (military jargon for the saying, "Situation Normal: All Fouled Up!"), a diversion, or even disinformation? We know that on July 7, Major Jesse Marcel was still up at the J. B. Foster ranch rooting through a field of strange wreckage and wouldn't return to the RAAF until the next morning. CIC Captain Sheridan Cavitt was with Marcel but would return to the RAAF with a load of wreckage in his jeep about midday. During the morning, however, the UFO impact site containing the only intact remains of the disintegrated UFO, plus three dead aliens and one "live one," was discovered north of Roswell. One could speculate that the call went to *Life* magazine the previous day (Sunday, July 6), before the Air Force realized what it was they were dealing with—a meteor? Then, when they understood the magnitude of what had come into their possession the next day (Monday, July 7), the Air Force changed its mind about having an outside entity such as *Life* magazine involved, and decided to lead them off course on a wild goose chase to get rid of them. If, however, the Air Force request to *Life* went out after the true nature of the crash was understood, the question then must be asked, Why? Why bring in an outside, civilian organization to perform a task when your own organization—the military—had boatloads of photographers within its ranks to do the job, including a score of photographers on the ground right in Roswell at the RAAF's own Third Photo Unit? On its face, this latter scenario doesn't make any sense at all, whereas the former one has a ring of plausibility to it, based upon the military's capacity for making mistakes or precipitous decisions with only rudimentary information in hand. Without anything else to draw upon, we might conclude that it had to have been a mistake, or that the Air Force had acted before it got its act together by deciding on a cover-up course. But we had an insider at our disposal who actually worked for *Life* magazine to offer perhaps a more informed opinion as to what might have transpired. What did she believe happened? Grant's wife, Karin, informed us that it is an axiom in the news business that the military and government use the media in the same way that the media uses them—"to try to get the others to do what you want them to do."[11] Regarding Roswell, she offered her opinion:

> Maybe the Feds knew all along that it was something more than a meteor, and what better way to deal with it than to invite the prestigious *Life* magazine to come in and take a look. You take them someplace near but not exactly to the spot and show the world that there's nothing there. Everybody is happy and relieved, and you can go about your own business. Additionally, you can keep all other media out by saying that *Life* had already been there and found nothing.

Mrs. Grant's observations presented us with another interesting angle, and quite frankly one that we wouldn't have thought of, from someone knowledgeable about the possible motives and tactics employed during the early stages of the Roswell cover-up. What better way to keep the mainstream media at bay than to tell them that one of their own—*Life* magazine—was already on the story? And it worked! There were no mainstream media stories anywhere about the crash beyond the initial wire stories that were all based on the military press releases of July 8 and July 9. That was it! The Roswell UFO story was a dead issue by July 10. Before he died, according to his wife, Allan Grant had abandoned his previously stated "innocent explanations" for what occurred so long ago and had come to the conclusion that it was no meteor he was chasing, but instead it had been "a cover for something big." Grant's wife said that whenever he talked to her about his Roswell trip in later years, "He always talked in terms of an 'unidentified flying object' that crashed near Roswell—not a meteor or anything else. That's what he came to believe it was."[12] And what about the gun? What did Allan Grant and Major Charles Phillips need to protect themselves from on their search? Rattlesnakes, an unfriendly meteor, or *something else*? Unfortunately, we do not have the answer to that question and, no doubt, never will.

6

HARASSED RANCHER SORRY HE TOLD: THE AFTERMATH OF A BALLOON RECOVERY—THE TRUE STORY

What is generally perceived by the novice is that after Mack Brazel turned most of the country on its ear with the story of the millennium, the "higher-ups" returned the world to reason and sanity with the most logical explanation they could muster. The press went for the bait, accepted the old switcheroo, and the poor, ignorant rancher went back to Corona a humbled man. Our national security was not an issue; rather, it was our overreaction to the fanciful claims of some wide-eyed intelligence officer at some military base in some one-horse town called Roswell. And, as General Ramey intimated, only those who over-imbibe see such things. Unfortunately, for those who tend to distrust officialdom on just about every issue but this one, the story does not end here. And sadly for Mack Brazel and those dear to him, it was only the beginning.

But first we will need to retrace just exactly what chain of events took place from the moment Brazel rightly or wrongly attempted to perform his civic duty by going to the authorities and report the crash of something beyond the realm of explanation. For all of his trouble, he became caught up in a web of malfeasance and cover-up that altered his life and that of his family even up to today. We will discover that in the government's world of sinister suppression and egregious intimidation, one does not just walk away unscathed after a Mogul balloon recovery, unless, as singer/actor Dwight Yoakam, in his portrayal of Mack Brazel in the highly acclaimed movie *Roswell* defiantly responded to his interrogators, "you're expecting some surprise balloon attack."[1]

There is absolutely no evidence that Mack Brazel *ever* spent an entire day driving all the way to Roswell, New Mexico, to report any type of weather balloon device—except for the one time on Sunday, July 6, 1947. This is a matter of indisputable fact. Certainly, all of the eyewitness testimony presented in this tome provides a wealth of circumstantial evidence arguing that Brazel was not alarmed over a mere balloon or anything else so prosaic. To the contrary, all of his actions and those of his neighbors suggest something truly out of the ordinary, which would eventually lead Brazel to fill up a couple of cardboard boxes with a selection of some of the debris and head off to Roswell to an ignominy that would haunt him for the rest of his life.

It is true that whatever Brazel brought to the attention of Chaves County Sheriff George Wilcox impressed him enough that he immediately dispatched two of his deputies from Roswell to it check out. What clearly elevated the entire affair to a national level of interest was the fateful phone conversation with KGFL's Frank Joyce. At that moment, the press learned of the whole dramatic affair, and if not for the military intervention, this book would be seventy-five years late, and obsolete. To Joyce's everlasting regret, the very worst piece of advice he provided the unfortunate rancher was to report the crash to the RAAF. No strangers to top secret tests in their immediate area, Joyce and Sheriff Wilcox thought, what else could it be?

Picture the following situation: It is a holiday weekend, and some rancher wanders into your town with an outlandish story about wreckage from a flying saucer strewn across the desert floor. Next, the local media gets wind of the story, and in short order the most elite unit of the U.S. military, in charge of our one and only nuclear bomb wing, is alerted to look into it. The rancher has quite a tall tale, but there's more: He has brought in physical proof. Point: Absolutely nothing about the physical remains of a Mogul balloon would have suggested it was part of a top secret project. Should we then assume that if the "evidence" were simply remnants from a weather balloon—neoprene rubber, wooden sticks, flimsy reflective foil, masking tape, and bailing twine— that not only the head of intelligence from the base, Major Marcel, but also the head of the CIC, Captain Cavitt, would have been ordered by the base commander, Colonel Blanchard, to make the long drive back with the rancher and personally investigate it? Marcel would have been sent in the event it was one of ours, and Cavitt if it was one of "theirs." Neither man questioned his superior officer after examining the assortment of debris just brought into the sheriff's office.

Since the moment Brazel tipped off the radio station, Joyce's boss, Walt Whitmore Sr., wanted to get his hands on the main source. At some time on Monday, July 7, while the two intelligence officers went about their assigned mission, Whitmore arranged for

The former Roswell radio station KGFL is now a hair salon. (Photo courtesy of Tom Carey.)

someone to grab Brazel at the ranch and bring him back down to Roswell. By that time, Whitmore surmised that the cowboy was about to become a heavily sought-after man. But where to hide him? Why, Whitmore's home— what better place? That very evening the station minority owner and news-man, "Jud" Roberts, conducted a wire-recorded interview with Brazel. Having already signed off broadcasting for the day, KGFL would have to wait until morning to break the biggest story in the history of the network.[2]

Unfortunately, time was not in their favor, and Brazel was quickly becoming a fugitive. Marcel and Cavitt were about to report back to headquarters. And for some unknown reason, Marcel had sent Cavitt long ahead of him to report back to Blanchard. According to Marcel, they couldn't identify the material.

As part of the Army's urgent attempt to get control of the entire situation, they grabbed Brazel and his recorded testimony first thing the next morn-ing. It was now Tuesday, July 8—just hours before posting the famous press release. It was all so contrived and calculated. As Brazel was thoroughly inter-rogated at the base, Marcel was ordered to report to "higher officials," and Blanchard was about to announce that he was conveniently going on leave after the holiday weekend. Now, with all of the principals out of the picture, the stage was set for it all to be rationalized away. By late that afternoon, Gen-eral Ramey denigrated Blanchard, Marcel, and some "old West cowboy" who was probably just looking for attention. "It's just a weather balloon!" he said.[3]

❖✦❖

Clearly, the story should have ended there—especially for Mack Brazel. But we would be remiss not to consider why the military found it necessary to take him early the next day to the *Roswell Daily Record* to recant the entire sordid tale. Surely Ramey's *official* explanation should have trumped that of any lowly civilian. Or was it part of a full-scale damage control program to further promulgate the government's stand on flying saucers—a preemptive strike, so to speak. Too many people, including the press, knew the true story. And up until the Fort Worth press conference, nobody was talking about any weather balloon—except for General Ramey. Should we believe that Brazel voluntarily made himself into a public spectacle? He had just spent the past *five days* trying to simply get someone to explain and clean up all that mess on a ranch he was hired to supervise. Everyone offered advice, but no one had any answers—not his neighbors, the state police, the sheriff's department, intelligence officers at the base, nor its commander. No one. Yet the Army was still compelled to hold on to Brazel and escort him with MPs to all the media outlets throughout Roswell *the day after the balloon explanation.*[4] And when they completed their derisive manipulation of the media, what did they do with the ranch foreman? They returned him back to the base for another *three days!* It is quite evident that this civil abuse of Brazel was due to the fact that he had witnessed more than just unusual wreckage. He had seen more— something that could hardly be explained away as a mundane weather balloon, or even an experimental vehicle made of some new exotic material.

During this disruption, Brazel had no hired hands to complete the ranch's chores. Cattle and sheep had to be fed and watered, and horses needed tending. The troops assigned to the special cleanup operation were oblivious to such trivial concerns, but Brazel's two older sons, after hearing about their father's disappearance, made their way to take over the ranch affairs.[5] Unfortunately, son Paul made the error of arriving first—during the military's occupation. Paul was a rancher in Texas at that time and had traveled some distance to help out his father—or at least try. To our frustration, as with so many others who saw too much, Paul would never discuss the situation with us. He always told us he had absolutely nothing to say. Finally, just before he passed away from cancer in 1995, he confessed one important concern as a rancher to his nephew Joe: "You know what always riled me even up to this day?" he asked. "Every time I tried to get to the main ranch house (ten miles from the debris field) to water the horses in all that summer heat, the damn Army forced me off the ranch. I tried again the next day and they still threw me off the property. I was sure they did nothing for any of the animals."[6]

Bill would fare somewhat better when he and his wife, Shirley, showed up from their home in Albuquerque just as the troops headed back to Roswell with all the remaining evidence.[7] They quickly went about the task of returning things back to normal. But such was not the case for his father when he unexpectedly showed up just as mysteriously as he had disappeared. Word was that

Mack Brazel, standing between his two eldest sons, Paul (left) and Bill. This photo was taken in the 1950s when Mack was working as a security officer in Tularosa, New Mexico. (Photo courtesy of Mrs. Fawn Fritz.)

he was flown back in a small plane. No apologies, no regrets for all of the mental and physical mishandling.[8] He was bitter and humiliated, and his good-natured spirit was broken, seldom to be seen again. Daughter Bessie would be taken aside by her dad and warned, "Don't believe everything you read about me in the papers. The government is going to use me to keep something secret!"[9] His son Bill would press him. "It's better that you don't know," Mack told him. Bill would persist. Cutting him off completely, his dad would slam the door: "You don't want to know."[10]

At this juncture we need to reiterate: Would all this have been necessary if the precept of a weather balloon were sufficient? Let's assume for the moment that it was still part of a top secret project—say, Mogul. Remember, the authorities were not looking for any missing test launch. No chase planes were patrolling the area. And still Brazel was subjected to the following violations of his civil rights:

1. Being physically abducted by the U.S. Army from the private residence of Walt Whitmore, along with a First Amendment violation for the confiscation of the wire recording from the media.

2. Being physically detained at the RAAF for up to five days at the base "guest house" for questioning. He could identify all his neighbors, yes, but what could Brazel possibly tell them about a top secret Mogul balloon? There is absolutely no evidence that he was ever accused of being a spy for a foreign government.

3. Being confined at the Roswell base for five full days without the benefit of due process. In fact, he was not permitted to make *any* phone calls—not even to his wife.

4. Being forced to undergo a full Army physical examination—for being exposed to balloon parts? Or was it a full-body search for the same? Brazel would later complain that he felt very degraded by this indignity. He would later complain upon his return home that he was kept up all hours of the night and asked the same questions over and over again. It has been suggested, based on the testimony of newsman Frank Joyce, who claimed similar treatment by the authorities, that Brazel was isolated on the final day of confinement and subjected to subtle brainwashing in a final attempt to silence him.[11]

In summation, it is abundantly clear that whatever Mack Brazel saw, it *was* definitely a threat to our national security. At least the Army's actions to pressure and coerce him into submission smack of strong-arm tactics generally reserved for captured enemies of the state—hardly the treatment of a mistaken man. The question persists: Does a Mogul weather balloon merit such measures?

For those who still insist on believing the official explanation, Mogul was officially declassified in 1972, but the primary objective was already clearly known in 1947. Mogul eventually became just another obsolete government test that had outlasted its usefulness. And still the Brazels remained suspect and under surveillance by agents of the U.S. government—possibly another reason for Mack's total silence about the incident.

The RAAF Guest House stood near the main entrance to the old base as seen in a 2008 photo. It is now gone. (Photo courtesy of Tom Carey.)

True to his word, whether out of loyalty to his country or just fear and deep concern for the safety and well-being of his family, Brazel never did talk. In fact, he went out of his way to avoid any conversation about this bleak time in his life. Hired workers such as Ernest Lueras remembered just how much Mack's demeanor had changed: "There was one particular time I rode along with him down to Tularosa [from Corona, three hours]," Lueras said. "This was right after he got into all that trouble with the Army. He didn't say anything. I tried to strike up a conversation. Not a word was said for the entire time I was with him. [The military] really messed him up."[12] "My dad was never the same," expressed Bill with a look of melancholy on his face.

Possibly driven by lack of resolution, his son continued to seek out some answers, eventually salvaging enough evidence to fill a cigar box. After heavy rains, Bill had known enough to return and check for anything missed by the military cleanup. But this time there would be no subsequent drive to Roswell.

It was something he regretted for the rest of his life—the day in 1949 when Bill turned over the cigar box to a Captain Armstrong and three NCOs[13] the morning after he had simply acknowledged "finding a few scraps" at Wade's Bar the evening before.[14] Whether out of shame or fear, Bill always described the encounter as rather straightforward and routine, but Fawn, to her everlasting horror, witnessed much more. While her father was leading the military intruders to the pasture, others, on a more circuitous mission, arrived at the ranch house. "There were six soldiers, who came right into the house," Fawn said. "They pulled drawers from dressers, emptied closets, and proceeded to pry up floorboards in Dad's bedroom. They completely trashed the house." It looked to be more of a warning than any methodical search. "From the house they went to the cattle shed and started to slit open each feedbag and let it pour over the ground," Fawn said. "They even emptied a water holding tank." They didn't say a word, and as fast as they had stormed the ranch, they fled like thieves, leaving the location pillaged and violated. Did anyone in authority even offer to pay for all the damage? we asked. The pain was still evident in Fawn's response: "How could this happen? How could this ever happen in this country? It was beyond belief. They made Dad swear not to say anything. First it was Granddad, then Dad." Who would be next?[15]

Sergeant Bert Schulz was a B-29 electrician assigned to the 393rd Bomb Squadron at Roswell Air Field (as it was renamed during the transition period of September 1947 to January 1948 from the RAAF to Walker Air Force Base, when the Air Force became a separate branch of the military) in December of 1947, just five months after the incident. Aside from the occasional scuttlebutt about all the activity at the "big hangar" a few months before, Schulz described a much more disturbing element, an element that soured him on the military, causing him to conclude that the ends don't necessarily justify the means. "There was still a lot of talk about the MPs harassing civilians over the affair. . . ," commented Schulz. "The MPs got pretty rough with some of the ranchers out there, and they were bragging about it!"[16]

Getting pretty rough with people was just the style of an Army Air Forces officer by the name of Hunter G. Penn. An associate of ours in the ongoing Roswell investigation, Anthony Bragalia, utilizing sophisticated investigative techniques, in 2008 located and interviewed the foster-daughter of retired Air Force Colonel Hunter G. Penn (now deceased). During World War II, Penn was the bombardier on a bomber in the 303rd Bomb Group known appropriately as Hell's Angels. After the war, she said that he was "associated" with Wright Field in the summer of 1947, and that when she was a teenager her father told her that he had undertaken a deadly serious assignment back in the summer of 1947. He explained to her that when he was at Wright Field, he was ordered to "visit places around Roswell, New Mexico," where a UFO had crashed. She told Bragalia that her father was tasked to (putting it euphemistically) "help manage civilian-military affairs after the crash." He was to ensure that an "information blackout" regarding the event was put—and kept—in force. Put bluntly, he was to question those who might know something about the crash, especially the little bodies, and "make sure that they did not talk." He was ordered to concentrate on ranchers, farmers, and "simple types" residing in the outlying areas who may have seen something, using intimidation and threats (with a weapon) to instill fear and compliance. The matter was so important that, if he could not persuade people who were not compliant, her father was to use physical force, including weaponry, to enforce silence. He confirmed to his daughter that the crash near Roswell was extraterrestrial and that they were concerned at the time about unknown dangers or problems that might be in the offing. And, echoing the words of former RAAF base adjutant Major Patrick Saunders (see Chapter

23 on deathbed confessions), her father told her that they didn't know where "they" were from, or what "their" intentions were.[17]

It appears that "in the heat of battle" at the time that the Roswell Incident was in progress during the first two weeks of July 1947, the military powers that be decided to employ "friendlies"—authority figures known to the townsfolk in Roswell and the surrounding environs (see Chapter 18)—in the personages of the local sheriff and the RAAF base security liaison officer to the City of Roswell to try to silence the local citizenry. But in the weeks and months immediately following the incident, after the initial fervor had died down, the question of how to maintain and enforce the silence of those who knew the truth became a paramount issue. Apparently, it was felt that appeals of national security and patriotism would not be enough to achieve the desired result of complete silence, especially on the part of the outlying ranchers, in time. In the military mindset of the time, the answer was to "Put the fear of God in them!" Enter Hunter Penn. It makes complete sense that they sent a stranger such as Penn to Roswell to do the really "dirty stuff"—to intimidate or kill. An officer from the RAAF would not have been used in such a long-term, systematic way to commit physical violence on local civilians. Such an officer would be known to the community and could be identified and held to account. An outsider, on the other hand, would be infinitely more preferable: someone who has no compunction about using whatever means necessary to get the job done, someone with no ties to the town, someone who would return to somewhere else when the job was done, someone like Hunter Penn.

The obvious issue here has to do with the urgency with which the military acted. Keep in mind that Project Mogul was a top secret program—*the project, not the materials!* Once again, the elements that composed a Mogul device were conventional and identifiable to everyone. And when it is broken up and scattered on the ground in front of you, there is nothing top secret about it. A cigar box of balloon scraps would hardly warrant the official reaction Bill Brazel received. He was not in possession of Mogul secret documents; all he ever admitted to was discovering pieces from the debris field, which the military wishfully assumed had been sufficiently vacuumed. Evidently, they were wrong. Clearly, the military was in a desperate search for something much more sensitive than balloon fragments. And most shocking of all, this entire incident happened *two years after the original crash*, which portends the true seriousness of this situation and explains why the Brazels were still being watched.

Months after returning home in 1960 after serving in the Navy, Mack's youngest son Vernon would disappear. Mack himself would die from a massive heart attack in 1963. Bill's son (Mack's grandson), William R. Brazel, was shot to death while hunting with two companions in 1964, and one other hunter was also killed by a second bullet. State police concluded that *both* were shot accidentally. Is it any wonder that Mack Brazel was sorry he ever told anyone back in 1947?

Former real estate salesman Howard Scoggin of Las Cruces, New Mexico, described a 1959 encounter with Mack Brazel. Scoggin had gone to a local restaurant for lunch with a friend who pointed out Brazel sitting alone at another table. Against his friend's better advice, Scoggin got up and approached the unsuspecting Brazel and asked him about the 1947 incident. Without saying a word, the rancher clenched his fist tightly, grimaced, and slowly rose out of his chair. Fearing for his personal safety, the surprised and now wary Scoggin backed away while Brazel slowly stalked past him and out of the restaurant, leaving his food on the table. "It was like watching one of those werewolf movies when Lon Chaney turns into the hairy monster," recalled Scoggin.[18]

KGFL minority owner Bob Wolf just happened to run into Mack at a festival in Corona mere months before he died. It was some years since the aging cowboy had passed through the area, so Wolf took advantage of the opportunity and brought up 1947 while exchanging social niceties. Brazel's entire attitude immediately changed. "He looked as though he had seen a ghost," described Wolf. "Those people will kill you if I tell you what I know!" said the rancher. He stormed away and unobtrusively slipped out from the gathering. Wolf would never see him again.[19]

On only one occasion did Fawn Fritz ever overhear her grandfather comment about the incident. "It was not uncommon for granddad to come riding up to the ranch house with an injured lamb cradled in his lap," she told us. "He treated all people and animals with respect and kindness. That is why it didn't surprise me when he softly spoke of 'those poor unfortunate creatures' back in 1947." But just as quickly he cautioned, "It wasn't anything anyone would ever want to see. Thank God you didn't!"[20]

7

"I SHOULD HAVE BURIED
THAT THING!"

Lincoln County, New Mexico, circa 1947: undeveloped, high-desert range-land unchanged for hundreds of years. Historically, ranchers pass down their vast landholdings from generation to generation. Families of cattle and sheep herders drove their livestock over miles of barren territory at as great a financial risk as most Wall Street investors. Cattlemen such as Proctor, Sultemeier, Strickland, Martin, Eppers, Pierce, Boswell, and McKnight had lifetime stakes in a profession that is acknowledged to have been one of the main influences of westward expansion during the 1800s. A celebrated life-style, long depicted in novels and movies, was in reality a demanding, rugged, harsh, seven-days-a-week occupation resulting in cracked-leather skin, blistered hands, saddle sores, and dealing with stubborn animals. Added to this were extended drought conditions, which assured a daily diet of trail dust and biting flies.

Mack Brazel was unique to the rest of these hardened ranch owners who were his neighbors back in the '40s. As anyone familiar with the Roswell Incident knows, Brazel was not the owner of the ranch or livestock he oversaw during those times. As ranch foreman, he nonetheless had an even greater responsibility for the daily management and operation of affairs. And unlike all the surrounding cowboys who participated in roundups, branding, bargaining wool prices, and selling cattle by the pound to local stockyards, Brazel's wasn't a vested interest. Rather, he answered to his boss, or, in his case, twin brothers from the vast plains of Texas, H. S. and J. B. (Jap) Foster. Well known throughout the Southwest, the Foster brothers operated vast ranches in not only New Mexico but also Midland and Kent, Texas. According to a great grandson, Cory Derek, the brothers also "made a fortune off of oil." Supposedly, many of the mineral rights the Fosters acquired were transacted immediately after 1947, and Derek is convinced that his relatives were paid off for their silence.[1] If this is true, then it would lend credence to the notion that Brazel also did the responsible duty and contacted the owners before reporting the crash to the authorities.

This was in fact confirmed by Geraldine Perkins, who ran a grocery store in Corona, which possessed one of the only two telephones in the town. According to her, Brazel used the phone in her store to make the call.[2] And just as it has been rumored for many years that Brazel was paid off for his eventual cooperation, so has it been suggested regarding the Fosters. After all, it was their property that surrounded the land where the crash took place. Access to their ranch was restricted during the recovery operation, and large areas of private property were of limited use for an extended period of time. Workdays and profits were lost, and, even if the Fosters missed the frantic call for help from Brazel, they would have certainly heard or read about it due to all the national media coverage.

What exactly did Mack Brazel earnestly tell his superiors? What has the Foster family maintained that crashed on their ranch for seventy-five years? Until his dying day, Brazel remained convinced that what crashed on the ranch was indeed an actual flying saucer. If it were any type of weather balloon, he would have been responsible for clearing it from the graze land, and that would have been the extent of it.[3] And even for the novice, rubber, wood, foil, tape, and string are all common elements used every day on a ranch. One might consider just what Brazel would have had to say to the Fosters to convince them otherwise.

It has often been our experience that any mention of the Roswell Incident by the press or the numerous cable documentaries that air yearly generate all forms of discussion. It was no less so within the families who were involved. Such was the case around 2001 when Cory Derek confronted his grandfather, who was H. S. Foster's son. Cory affectionately called him "Papa." Like so many other reluctant witnesses we have dealt with through the course of our investigation, Cory's grandfather was tight-lipped. "He was totally evasive about the whole subject, and the way he behaved was unlike him," described

Cory. So the matter was dropped, but when "Papa" passed away in 2005, Cory took the opportunity to bring it up again with his uncle. At first, the uncle was willing to speak about the event, but, as you will see momentarily, the proverbial Roswell monkey wrench would end the discussion. Initially, he informed his nephew that he was seventeen when the crash occurred and that he was working for the H. S. Foster ranch in Texas at the time. When asked directly if it was a flying saucer, he replied, "Those boys up there told me they were certain without a doubt that what they saw was a flying saucer." But next came the monkey wrench. Cory asked him, "Did they see bodies?" The uncle stammered, "Uh, uh . . . I can't remember." Lastly, he slammed the door with, "That is all I remember about that." Cory observed, "There was a certain gravity to his closing remark and the subject was never brought up again."[4]

Shedding additional light on the involvement of the Foster family is the daughter of J. B. Foster, JoAnn Purdie. She said, "My dad knew it was flying saucer and never changed his story. And just as the Army warned and threatened Mack Brazel, they did the same to him."[5] Foster's daughter recalled how, during the numerous times she encountered Brazel after 1947, he refused to discuss the incident. "He would state that it wasn't any weather balloon," added Purdie. She also agreed that Brazel definitely contacted her family about the crash and then observed how it changed the behavior of her father after he drove to his New Mexico ranch from Texas. "Whatever he saw or heard for himself, he too didn't want to talk about it after he returned home," said Purdie. To this day she still remains suspicious about the death of Mack Brazel in 1963 due to specific comments made to her by her dad. "I have absolutely no doubt he believed all the threats, and that they meant business," finished Purdie.[6]

Just as Major Jesse Marcel served as the fall guy for the military (detailed in Chapter 12), one could reason that Brazel was the proxy for the Fosters. After all, it was their ranch, and their property, and he was their employee. They could have been held accountable for whatever happened on one of their ranches. The argument could easily be made that, just as it was Marcel's picture plastered all over the newspapers with the spurious weather balloon, it was Brazel who was escorted by the military into the office of the *Roswell Daily Record* to retract his original contention. His own Waterloo would result in the highly controversial newspaper article of Wednesday, July

9, 1947, entitled, "Harassed Rancher Who Located 'Saucer' Sorry He Told about It." No one, including Mack Brazel, has ever described the crash debris as spelled out in this phony article. It remains inconceivable that, if we are to believe this report is accurate, it would generate a tenth of the reaction it did—from the outside ranches, the press, the military, and beyond. If it can be demonstrated that a weather balloon has ever spawned such a national security fiasco, we would love to learn about it. Instead, we are still asked to accept that Brazel sauntered into the newspaper office to report a bundle of rubber, foil, and sticks that, for some reason, everyone else believed to have been an actual flying saucer.

One last point of refutation: the last line in the *Record* article quotes Brazel defiantly stating, "It was not any type of weather balloon. That, I'm quite sure of." Even among Roswell advocates, the consensus remains that the rancher still managed to blurt out the truth despite all the military coercion. Their dilemma is that, even if correct, the Army escort still would have had sufficient time to prevent the paper from printing this inconsistent remark. Stated more simply, this too was part of the cover-up script. Brazel's final testimony regarding the matter had to demonstrate some level of confusion and ignorance on his part. He had no intention of misleading everyone about something that inevitably was identified as just a weather balloon. The final word had to leave the general public with the lasting impression that it was only Brazel who didn't recognize such a common device and mistook it to be something truly out of the ordinary.

Mack Brazel's late daughter, Bessie Brazel Schreiber, was initially one of the Air Force's "star witnesses" in their July 1994 Project Mogul report by supporting the Air Force's contention that her father had recovered a "high altitude" balloon in the first week of July 1947.[7] In recent years before she died, however, she recanted that position and conceded, "It was another occurrence altogether. I had helped my dad gather up weather balloons on a number of occasions. I have come to the conclusion that what my dad found back at that time was something else altogether."[8] It is accepted that she and her brother Vernon were at the ranch at the time of the incident, but the ranch house was almost ten miles from the debris field, and we have every reason to believe that Vernon eventually saw as much as his father. Still, there remains no reason to believe that Bessie saw anything out of the ordinary. As her older brother Bill once charged, "She wasn't even there!"[9] He was referring to the wreckage site. The question remains: Did Mack Brazel drop his children off at their true home in Tularosa on his way into Roswell on Sunday, July 6, 1947, with a foreboding premonition of what was to come? Bessie believed so. "Dad told us that the authorities would probably use him to hide all this from the people." But, she added his one caveat, "Don't believe everything you read in the papers about your dad over the next few days."[10] All of this would tend to emphasize just how overwhelming the whole situation was. Mack Brazel was about to change the course of history—for better or worse.

In 1999, the late daughter of Mack Brazel, Bessie Brazel Schreiber, gave the authors her final interview regarding the 1947 Roswell UFO crash events. (Photo courtesy of Tom Carey.)

JoAnn Purdie described one last phone conversation her father had with Mack Brazel years after the whole earthshaking affair was but a dim memory. The aging rancher complained to her dad that "I should have buried that thing!" She closed, "In every way the entire ordeal had ruined his life."[11] But truth has a nasty habit of rising to the surface, no matter how deep you bury it.

8

"NOTHING MADE ON THIS EARTH"

I t is now seventy-five years after the event, and it is still unclear whether RAAF Major Marcel knew that the mangled remains of a weather balloon and radar kite were part of the secured cargo he escorted aboard a B-29 bomber to a meeting with General Roger Ramey in Fort Worth, Texas, on July 8, 1947. One thing, however, is certain: until the day he died in May of 1986, Marcel swore that what he personally recovered in the desert north of Roswell was not what was displayed for *Fort Worth Star-Telegram* news photographer James B. Johnson in General Ramey's office on July 8. And, as if to add insult to injury, after Marcel was instructed to pose with the substituted weather device, Ramey also ordered the major not to say a word to any of the press waiting outside the room.[1]

As the chief intelligence officer of the 509th Bomb Group at Roswell Army Air Field, Marcel likely found it strange that he was all but completely eliminated from the press conference that followed his photo op. Marcel was kept in seclusion for the next twenty-four hours; he finally was allowed to return to Roswell on the evening of Wednesday, July 9. The B-29 that carried Marcel from Fort Worth back to Roswell was nicknamed Straight Flush and bore the tail number 447301. It was piloted by Captain Frederick Ewing.[2]

Major Marcel returned to his home in Roswell just one day after he had displayed some of the true wreckage to his wife, "Vi" (short for Viaud), and son, Jesse Jr. Upon his return, Marcel informed both of them that he was no longer able to talk about it with them. But that didn't prevent him from trying to get some answers back at the Roswell base.[3]

The very next morning, July 10, Jesse Marcel arrived at his office and confronted the officer who had accompanied him to the Foster ranch with Mack Brazel to first investigate the crash—the head of counterintelligence for the Counter Intelligence Corps (CIC), Captain Sheridan Cavitt.

JM: I want to see the report of what all happened here while I was in Fort Worth.

SC: What report? I don't know what you're talking about.

JM: I outrank you!

SC: Sorry. I take my orders from Washington. If you don't like it, you can take it up with them.

On that brusque note, the CIC officer put an abrupt end to the debate.[4]

It should be pointed out that CIC NCO Lewis "Bill" Rickett was also present at the heated exchange. To him, the preceding discussion was out of character for both men; the Marcels and the Cavitts were good friends.[5] Furthermore, the captain's wife, Mary Cavitt, told a most interesting story. According to her, within a few nights of Jesse's return from his special flight to Fort Worth, the two couples got together for their weekly game of bridge. But this night was different. The wives remained in the main room while the two husbands toiled over the stove in the kitchen. As Mary described it, "The men turned up the heat as high as it would go, and it still had no effect."[6] That's because the focus of their efforts was no simple pot of boiling water—it was a piece of the *real* crash debris. Mrs. Cavitt claimed this rather astonishing scenario ended when "Cav [her husband] reminded Jess that the material was classified 'top secret' and he had better get rid of it." According to her, "The two men went out onto the patio" with the nigh-indestructible material. Moments later, when they returned, it was gone. "It never came up again." Loyalty and security oaths had prevailed.[7]

During a radio telephone interview with KOAT in Albuquerque in 1985, Marcel ended his recollections by adding this caveat: "I haven't told everything." With Linda Corley, who interviewed him for a school project just before he passed away in 1986, Marcel was more explicit, "There is a hell of a lot I haven't said. I can't for the sake of my country. I'm a trained intelligence officer. Once you do that, it's with you for life." One would think that the head of intelligence at Roswell would have been privy to all details regarding the incident. True, he was removed from the scene for the better part of two days during the most crucial part of the recovery operation, but if indeed there were bodies involved, how could Marcel not have known the truth? Even today, his son Jesse Jr. maintains that his father never uttered a word about such things. Others might disagree.

Our investigation has concluded that Major Marcel *had* to know about the alien bodies that were recovered from the crash—not secondhand by hearing about them from others in the chain of command, but firsthand from actually seeing them himself. If you accept that Mack Brazel found alien bodies *someplace else* on his ranch (detailed in Chapter 6), as he told the KGFL radio announcer when he first came into Roswell on Sunday, July 6—which we do—then you have to believe that Brazel not only told Major Marcel and Captain Cavitt about the bodies, but also showed them to the officers when they followed Brazel out to his ranch in Lincoln County to look at the wreckage on July 6, 7, and 8. That this may have been the case is suggested in an interview we conducted with former Tech Sergeant Herschel Grice in 2002. Grice was a ground maintenance crew chief in the 715th Bomb Squadron on the Roswell base in 1947. He was also a member of Marcel's intelligence team working within the 715th. Grice said that he knew Marcel well and that Marcel was a good officer and a "straight arrow" as a person. He also said that Marcel once told him about seeing the alien bodies. Grice couldn't remember any of the details other than to say that Marcel had referred to them as "white, rubbery figures."[8] Captain Darwin E. Rasmussen had been a flight operations officer with the 715th Bomb Squadron of the 509th Bomb Group in 1947. He later told family members that the Roswell Incident was true, and that four bodies had been recovered. Rasmussen's cousin, Elaine Vegh, said that she personally overheard him tell her father that he had no doubt that UFOs were real because he had helped to retrieve the bodies from the one that crashed at Roswell.[9] She also overhead her cousin tell her father that Major Jesse Marcel helped with the retrieval of the bodies.[10]

Sue Marcel Methane is a family member of the Houma, Louisiana, Marcels, from where Jesse Marcel hailed, and to where he returned when he left the Air Force. In discussing her famous relative with us in a telephone interview in 2002, she said that she had a chance to talk to him shortly before his death in 1986. She did not recall how or why the conversation turned to the subject of the Roswell Incident, but it did. Methane said that the one thing that stuck with her from that conversation was his description of the dead aliens. "He referred to them as 'white powdery figures.' I can't picture them, but that's what he called them." In 1998, Jesse Marcel's cousin, Nelson Marcel, was quoted in the February 15, 1998, edition of the Houma *Courier* newspaper as saying that Jesse Sr. told him that he had seen several "pygmy" alien bodies among the debris. Since the publication of our book in 2007, other Marcel family members have come forward. Hayes Marcel, a seventy-seven-year-old nephew of Jesse Marcel Sr. through marriage, told a *Houma Today* reporter that "Those must have been aliens in that spaceship that crashed [in 1947]. Jesse said he saw them." He heard this, not directly from Jesse himself, whom he did not know that well, but from another Marcel family relative, the aforementioned Nelson Marcel, who was Jesse's cousin (now deceased). According to Hayes Marcel, his uncle Nelson had discussed the 1947 Roswell events extensively with Jesse, who told him that, in addition to seeing the alien bodies, he had kept a piece of "unbreakable alien metal" from the wreckage for himself. Hayes Marcel also attested to the Marcel family's frustration with the government's continued "stonewalling" about Roswell since Jesse Sr. publicly broke his silence on the matter in 1978. Since then, many members of the extended Marcel family have hoped that our government would confirm the truth of Jesse Marcel's public statements. "I hope that it does before I check out," Hayes Marcel said. "All my relatives (hoped) too, before they died. The government knows a lot it's not letting out."[11]

The statements of Herschel Grice, Sue Marcel Methane, Nelson Marcel, and Hayes Marcel, rightly or wrongly, call into question the statements made by Jesse Marcel Jr., son of the RAAF intelligence officer, about what his father did and did not tell him about Roswell. For many years, Jesse Jr. has publicly stated that his father, who passed away in 1986, never mentioned to him that he had seen alien bodies. He still says that. To resolve this apparent conundrum, we are left with only two possibilities. Either Jesse Marcel Sr., in fact never told his son, who lives in Montana, about the bodies, while telling more distant family relations down in Houma, Louisiana, about them—perhaps to protect his son?—or, he *did* tell his son about them at some point, in which case Jesse Jr. hasn't given us "the rest of the story," for reasons known only to him.

MARCEL AT A GLANCE

Major Jesse A. Marcel—in charge of Army Air Forces security and intelligence briefings at Kwajelein Base in the Pacific (command center for Operation Crossroads, which test-fired two atomic bombs).

Maj. Marcel—honored with the Air Medal for operational combat missions and the Soldier's Medal for meritorious achievement in military operations during World War II.

Maj. Marcel—recipient of three commendations including one from Gen. Ramey.

Maj. Marcel—received a diploma in 1945 from the Army Air Forces Training Command in radar technology at Langley, Virginia, and was highly trained in radar tracking materials and equipment including rawin targets and ML-307 reflectors, which were the dominant parts of Project Mogul balloon devices.

Maj. Marcel—head intelligence officer, A-2, of a select squadron of people in charge of the atomic bomb in 1947.

Maj. Marcel—remained the head of intelligence at Roswell for another year and was then promoted to lieutenant colonel in the Air Force Reserve with recommendations from both Blanchard and Ramey.

Maj. Marcel—transferred to Washington, DC, where he was made the SAC chief of a foreign technology intelligence division.

Maj. Marcel—at the Pentagon's insistence, was assigned to the top secret Special Weapons Project.

Maj. Marcel—prestigiously honored with fifteen awards for combat credit and fifteen more decorations, including the Bronze Service Star, for his dedication and devotion to his country.

Maj. Marcel—scapegoat, whom the U.S. Army would have us believe was so incompetent that he couldn't identify the rubber, wooden sticks, foil, tape, and string that comprised a very common weather device.

Military scoffers have suggested that the reason Marcel was never publicly reprimanded or even demoted for attempting such a canard was to avoid attracting any additional embarrassment to the already humiliated Army. But this was far from an isolated internal affair. No one can credibly prove that Marcel, with the willing assistance of the public information officer, Walter Haut, cavalierly put out a press release without outside knowledge or permission. All of this we are asked to just chalk up to poor judgment and overreaction on the part of two overly zealous officers?

Colonel Blanchard and the brass all the way to Washington had wreckage in their possession since the day rancher Mack Brazel brought it into town— *two full days before the announcement went out!* Every action Marcel took in the entire Roswell affair was following the orders of his superiors. The RAAF publicly stated it had a flying saucer in their possession—not Marcel. And after the public spectacle subsided, clearly the higher-ups were much more confident with Marcel than the general public. But then again, the public accepted the balloon story. Apparently, the military knew otherwise.

During the last few years of Marcel's life, as he gradually struggled more and more with emphysema, he bravely tried to tell the world the truth about what he saw and handled out on that patch of desert property so many years before. "It was nothing we had ever seen before," Marcel stated on the record. "It was not an aircraft of any kind; that I am sure of. We didn't know what it was. It was *nothing made on this Earth.*"[12]

9

THE SENATOR AND THE ALIENS:
"GET ME THE HELL OUT OF HERE!"

On Monday, July 7, 1947, New Mexico's newly elected, thirty-two-year-old lieutenant governor, Joseph "Little Joe" Montoya, was in the sleepy desert town of Roswell. There are no written records that will confirm the exact whereabouts of Little Joe that day,[1] but thanks to the long-term memories of a few *Montoyistas* (young political supporters of Montoya), we know for a fact that Montoya was in Roswell. And the town of Roswell—especially the air base just to the south—was anything but sleepy that day.

The brothers who would become lifelong friends of Joe Montoya, Ruben and Pete Anaya, lived next door to one another on Albuquerque Street in the south end of Roswell in 1947. Both were active in local Democratic politics, and both were card-carrying *Montoyistas*. Though U.S. Senator Dennis Chavez was the highest-ranking elected Hispanic politician in New Mexico at the time, many in the state's Hispanic community felt that Chavez had become too "establishment" from his years in Washington and, as a result, did not adequately represent their interests and concerns. "Little Joe," as they affectionately referred to the then lieutenant governor, was their man, and the rest is history. Almost.

It is not entirely clear how Joseph Montoya came to be on base at the Roswell Army Air Field on the exact day and moment that wreckage from the downed UFO began arriving there. Also arriving was the first set of "little bodies," including a possible survivor from the crash site just north of town. The site had been accidentally discovered thirty-five miles northwest of Roswell in Chaves County earlier that day by a group of civilian archaeologists who telephoned the sheriff's office and fire department in Roswell from a service station in the nearby hamlet of Mesa. This site was closer to Roswell than the debris field site in Lincoln County that the RAAF's Major Jesse Marcel and Captain Sheridan Cavitt had gone to the previous day with sheep rancher Mack Brazel. Marcel and Cavitt were at this time still in the field and had not yet returned to Roswell.

The central depository for the retrieval operation, including the bodies, was Hangar P-3 (today known as Building 84), located along the flight line at the east end of the base proper. One account has Montoya being called to the air base in Roswell, as the highest-ranking state official in the area at the time, to view the wreckage and the bodies. (NOTE: It is now known that New Mexico Governor Tom Mabry suddenly took an "impromptu vacation to the mountains" at this time; he had been scheduled to publicly sign an Air Force Proclamation with the RAAF's Colonel Blanchard in Santa Fe on July 9, but the signing was deferred until July 14—after the Roswell events—with Joseph Montoya signing in place of the governor! It is also now known from President Harry Truman's daily appointment logs of the day that New Mexico's senior senator, Carl Hatch, requested an emergency meeting with the president on July 7; the requested meeting finally took

Joseph M. "Little Joe" Montoya was New Mexico's Lieutenant Governor in 1947 and later rose to prominence as a U.S. congressman and senator. (Photo courtesy of the U.S. Senate Historical Office.)

place on July 9.)[2] The other account, told to us by the Anaya brothers, put Montoya on the base with other public officials for a regularly scheduled special event—the dedication of a new airplane. After the dedication, according to the late Ruben Anaya, Montoya went over to the hangar area to greet some local, civilian Montoyistas who were working there and who wanted to meet him. He arrived in the vicinity of Hangar P-3 just as the first military vehicles were bearing down with wreckage and bodies from the crash site north of town. What then ensued was something that would cause Montoya to admonish his friends to never say anything at the threat of being called liars, something that would bring Senator Chavez and Chaves County Sheriff George Wilcox into the case to enforce the secret, and something that Montoya's family refuses to discuss by claiming that he was not involved.

Our investigation was first made aware of Ruben and Pete Anaya in 1991 during a book signing in Roswell for the then recently released *UFO Crash at Roswell* when Ruben Anaya's daughter introduced her father to coauthors Kevin Randle and Don Schmitt. It was in a number of interviews that we had with Ruben Anaya, Pete Anaya, and Pete's wife, Mary, in 1991, 1992, 1994, and 2002 that we learned the details of Joseph Montoya's extraordinary 1947 encounter. Ruben Anaya was also interviewed separately in 1993 by an anti-Roswell researcher, the late Karl T. Pflock, for his 2001 book, *Roswell: Inconvenient Facts and the Will to Believe*. The Anayas were also interviewed in 1997 by British researcher Tim Shawcross for his book of the same year, *The Roswell File*. The following narrative of the 1947 events involving Joseph Montoya represents a synthesis of the details gleaned from these interviews.

The first inkling that Ruben Anaya had that something was amiss was a panicked knock at his front door. It was his father informing him that Joe Montoya had just called him (neither Ruben nor Pete Anaya had a telephone) from the base to tell him to get Ruben to come out to the base as quickly as possible. Ruben Anaya went to his father's house and returned the call to find out what Montoya needed. Anaya recalled that Montoya sounded "real excited" and "panicky-sounding" over the phone, "like he had seen a fire or something." Talking as fast as he could in Spanish, Montoya told Anaya, "I'm at the big hangar. Get your car, Ruben, and pick me up. Get me the hell out of here! Hurry!" Returning home with alacrity, Ruben Anaya went next door to his brother's house to confer with Pete Anaya, who had been chatting with two fellow Montoyistas, Moses Burrola and Ralph Chaes. All four then piled into Ruben's car and headed for the base.

The Anaya car had no difficulty in passing through the main gate at the RAAF because Ruben worked on the base as a cook at the Officers' Club and as a recently discharged WWII veteran, was also a member of the NCO Club. Because of these affiliations, his car had an official base sticker prominently displayed, which the guards at the main gate recognized, so they waved the car on through. The base also had not yet gotten its act together in dealing with this unprecedented emergency and therefore had not yet gone into "lockdown mode."

Not wanting to be seen in this situation by anyone who would recognize him, Montoya had told Anaya not to drive near the base headquarters located in the center of the base because there were too many people there. Following these instructions, the Anayas finally arrived at the large water tower that still looms ominously over Hangar P-3, but could go no further, as their route was blocked. It was then they noticed the burgeoning excitement around the hangar, and that there were MPs as well as some city police controlling the entrance to it. Just then, the side door of the hangar opened and out sprinted Little Joe Montoya, "like a bat out of hell," according to Ruben Anaya. Montoya quickly got into the back seat of the car and exhorted, "Come on, let's go. Let's get the hell out of here!" Ruben Anaya

noticed that Montoya was very pale, "like he had seen a ghost or something," and was shaking uncontrollably. Anaya concluded that something must have frightened his friend severely. "He was very, very scared," Anaya would tell us years later. Pete Anaya would also attest to Montoya's strangely frightened appearance that day. Ruben then asked Montoya if he wanted to be driven to the Nickson Hotel, where Montoya normally stayed when he was in Roswell, but Montoya instead responded, "No. Just take me to your house. I need a drink bad." On the ride back, Montoya, at least initially, was still in a state of high excitement and anxiety as he kept shaking and twitching while rocking back and forth and

The old water tower where Lt. Gov. Montoya told the Anayas to meet him and "get me outta here!" still stands in front of Hangar P-3 [now Bldg. 84] in the background on the old base. (Photo courtesy of Tom Carey.)

muttering to himself, "They're not human! They're not human!" Later, he sat quietly, just staring out the window during the remainder of the drive back to town.

When they arrived at Pete Anaya's house, Montoya collapsed in a heap on the sofa. He was handed a small glass of scotch, which he quickly downed, but with no apparent effect. Montoya wanted something else. He was then given a bottle of Jim Beam, three-quarters full, which he then drank straight from the bottle in three big gulps, "boom, boom, boom," as Ruben would later describe it. They told Montoya to take it easy. "No. I've got to calm myself down. You are not going to believe what I've just seen. If you ever tell anyone, I'll call you a damned liar."

Still excited, Montoya launched into his account. "We don't know what it is," Montoya allowed. "There was a flying saucer. They say it moves like a platter—a plane without wings. Not a helicopter." Ruben Anaya revealed to us that when Montoya got excited, he tended to speak in Spanish; therefore, his "close encounter" was described to them mostly in Spanish—*un plato muy grande con una machina en la media* ("a big saucer with a machine in the middle"). "I don't know where it's from. It could be from the moon. We don't know what it is."

We don't know if Montoya actually saw the craft or was simply told about it by crews returning from the crash site. We suspect that it was the latter, because the recovery had just commenced and other eyewitnesses only described seeing pieces of wreckage being brought into the hangar at this time—not an intact vessel—along with something else. Montoya seemed to confirm this himself a little later by stating that he had seen wreckage being brought into the hangar when he was there, but that there was nothing that resembled an intact ship, just pieces of metal he didn't get close enough to examine.

Montoya wasn't finished. He then told the Anayas that he had also seen "four little men." He described how small they were, along with the stunning fact that "one was alive!"[3] There is a question as to the exact number of "little men" Montoya said he saw that day. Pete Anaya confirmed his brother's belief that Montoya had said there were four, but Pete's wife, who had been an interested bystander in the living room, thought he had mentioned only two. We believe that the correct total under discussion is four—three dead and one alive. Montoya described the beings as "short, only coming up to my chest." Ruben Anaya added that "Montoya stood up as best he could and held his hand up to his chest with his palm facing down to indicate how tall they were. Montoya was a little guy himself, which would mean that they were about 3½ feet tall." Continuing with Montoya's description, "(They were) skinny with big eyes shaped like tear-drops. (The) mouth was real small, like a knife-cut across a piece of wood, and they had large heads."

He then described for the Anayas the scene inside the hangar. Each of the little men, including the one that was alive, was stretched out on a table brought over from the mess hall and set up for that purpose. "I knew that one was alive, because I could hear it moaning." Montoya said that it was moving, with its knees up on the table, and that these, plus one hand, were also moving. "They were so skinny that they didn't look human," Montoya told them. As with the wreckage, Montoya could not get as close to the little men as he would have liked, this time due to the press of doctors and technicians around the tables. He got close enough, however, to see that their skin was a pale white and that they had no hair. Each wore a silvery, tight-fitting, one-piece flight suit. "From what I could see, they had four long, thin fingers [on each hand]. They had large [for their size], bald heads. I wasn't close enough to see what color their eyes were, but they were larger than normal." Just prior to his dash from the hangar, Montoya said that the little men were taken over to the base hospital. Exhausted from his ordeal, and exasperated at Ruben Anaya for asking so many questions, Montoya had reached the end of his tether. "I tell you, they're not from this world!" Upon hearing this, Ruben Anaya recalled, "We thought, 'This guy, he's out of whack.' That's when I shut up."

After finishing the recounting of his day's "not from this world" activities, Montoya finally began to relax. He lay back on the sofa and fell asleep, but it was a fitful sleep. He kept jerking and twitching himself awake, as if he were under a great deal of stress. After a while, he woke up and asked Ruben Anaya to call his brother Tom Montoya, who worked at the Nickson Hotel in downtown Roswell where he was staying, to come over and pick him up. Anaya, after placing the call from his father's house, arrived back about the same time that a car driven by another Montoyista, Donald Wortley, was pulling up to his brother Pete's house (Tom was unable to leave his duties at the hotel and had called Wortley to go to the Anayas' in his stead). Montoya came out of the house, and Wortley got out and helped him into the car. The group of loyal Montoyistas just stood there in a state of semi-shock, watching in awed silence, as the car sped away.

The following morning, Ruben and Pete Anaya drove to the Nickson Hotel to check on the lieutenant governor. A now more composed Montoya gave the brothers a little update as to what was going on: "Confidentially, they shipped everything to Texas, and those little guys are in the hospital." Montoya then reiterated for the Anayas his admonishment of the previous day, that if they talked to anyone about what happened at the base the day before, he would call them liars. With that, the Anayas left and went home. That evening, the Anayas had an unexpected visitor. Because the two Anaya brothers and their families lived next to each other, a tired and stressed Sheriff George Wilcox was thankful to have a "twofer" on his hands. The Anayas had no idea why the Chaves County sheriff's car would be stopping in front of Pete Anaya's house. Both families went outside and met Sheriff Wilcox on the front lawn. Wilcox had a message for them. Unknown to the Anayas, Wilcox had delivered the same message to certain other civilians in Roswell that day at the behest of the Army Air Forces. He told the Anayas in no uncertain terms that if they talked to anyone about what the lieu-tenant governor had told them, everybody, including the children, would be killed![4] After delivering his civil-rights–violating threat, Wilcox returned to his cruiser and drove off, leaving the stunned and outraged Anayas to contemplate their fate. On several subsequent occasions, Ruben Anaya said that he, his brother, and Moses Burrola did in fact try to discuss the episode with Montoya, who was by this time talking about much more than just calling them liars. "It's too dangerous to talk about. The FBI will do away with you," he said. Could that be what Wilcox meant? Anaya wasn't sure, but the warning was again clear. In a later discussion, Montoya repeated the warning: "If you talk about it, someone, maybe not the FBI, but someone in the government will get you."

A different appeal was tried on the Montoyistas by Joseph Montoya's longtime New Mexico adversary, Senator Dennis Chavez. Sometime after Montoya's experience in the hangar, as Ruben Anaya recalled it, Senator

Chavez summoned him, his brother Pete, Moses Burrola, and Ralph Chaes to the Nickson Hotel where he was staying to discuss something. After preliminaries, Chavez bluntly told them, "Joe Montoya is a damn liar! He didn't see anything. It was a secret project, and it could hurt us with Russia and Germany if word of it got out."

By the decade of the 1970s, when Joseph Montoya was a sitting U.S. senator, the "UFO crisis" of those early days had long passed. UFO sightings were still being reported, to be sure, but the Air Force was officially out of the UFO business, having closed its Project Blue Book in 1969.[5] And, out of fear of ridicule, there was no longer any talk of "crashed saucers" or "little green men." It was in this milieu that John Anaya, the son of Pete and Mary Anaya, joined the staff of Senator Montoya in Washington, D.C. We had an opportunity to talk to John Anaya at his parents' home in September 2002 when we were videotaping his parents for our Sci Fi Channel documentary, *The Roswell Crash: Startling New Evidence*. We asked him if he had known Little Joe Montoya. "Yes, I knew him well," he said. "I worked for him for ten years." We went right to it and followed up the first question with, "Did you ever ask him about what your parents are testifying about today?" "Yes, I did," he said. "I asked him if it was true. He said that it was, but that if I ever told anyone, he would deny it."

Ruben Anaya passed away in 2002. Moses Burrola died many years ago, but before he passed, he told his wife about that unforgettable day in July of 1947 when Little Joe Montoya came to Roswell. She confirmed for us that her late husband had indeed known Montoya and had been good friends with the Anayas. She could not remember any details of that day, however, only that something very unusual had happened. Ralph Chaes was never located or interviewed and is presumed to be long gone. In 2004, we received a tip from someone who called the International UFO Museum & Research Center (IUFOM&RC) in Roswell suggesting that we talk to a Roswell resident by the name of William "Bill" Glenn, the owner of Glenn's Furniture in Roswell. The caller said that Glenn had been Joe Montoya's personal pilot back in 1947 and might know something. Schmitt followed up on the lead by talking to Glenn, who informed Schmitt that he wasn't Montoya's pilot until much later, and that Montoya's pilot in 1947 was a fellow by the name of Donald Wortley, who was still alive and living in Roswell. According to Glenn, Wortley had been Montoya's driver, pilot, and all-around coordinator of Montoya's itinerary whenever Montoya was in Roswell.

We had always assumed that it was the Anaya brothers who had driven Montoya to the Nickson Hotel or back to the base to catch a flight after leaving Pete Anaya's house. And Karl Pflock, in his 2001 book, stated that it was Montoya's brother, Tom, who had picked him up and driven him back to the Nickson. Before approaching Wortley, we ran this new information by Pete Anaya, as Fred Willard's name was new to us and had never come up in any of the previous discussions with the Anayas about Joseph Montoya. "Yes. It was Donald Wortley. I had forgotten, but it was Fred who came and got Montoya that day." We then approached a friend of ours, the late Bruce Rhodes, who had worked at the UFO Museum in Roswell, and who had been by coincidence a friend of Wortley's, to try to arrange a meeting. Rhodes suggested that, because Wortley was known to have his morning coffee the same time every day at the midtown Denny's restaurant located not far from the UFO Museum, we could probably meet there. We gave Rhodes the go-ahead to set up a meeting.

As we waited with Bruce Rhodes in our booth at Denny's for Donald Wortley to show, it crossed our minds that maybe he wouldn't. After all, it wouldn't be a first. Rhodes had told Wortley only that we were going to ask him some general questions about the Roswell Incident, whether he knew anything about it or not. Joseph Montoya's name had not been mentioned as a possible topic of discussion. Sometimes that was enough, in our experience, to cause the Roswell-related affliction, *cold-feet-itis.* "Here he comes," said Rhodes, who waved him over to join us.

After exchanging pleasantries and a few softball questions about Roswell and the 1947 incident, we got down to business—in a pleasant way, of course. Wortley acknowledged that he had been Montoya's coordinator and personal driver whenever Montoya was in Roswell, and that he later became Montoya's personal pilot, as well. Most importantly, he acknowledged being Montoya's driver in 1947. As our questioning began focusing on his relationship with Joseph Montoya, we noticed that Wortley was becoming noticeably agitated and less and less talkative. We then dropped the big one on him and asked if knew anything about Montoya's "close encounter" out at the base in 1947. "No, I don't. Never heard that one." "He never told you about it? You were his driver in 1947. Wasn't it you who picked him up at Pete Anaya's house and drove him back to the Nickson after his episode out at the hangar?" At this line of questioning, Wortley's entire demeanor began to change. He began to appear unstable in his chair. His lower lip began to quiver uncontrollably, and his right hand started to tremble as he tried to stir his coffee. His face turned visibly pale as he lowered his head to avoid our eyes by staring at his coffee cup. We were looking at someone whose life appeared to be passing right before him (and us) as we sat there. Wortley never answered those questions or spoke coherently again that day. We left the restaurant feeling truly sorry for him, but also feeling that our questions had indeed been answered.

10

COVERING UP THE BIGGEST STORY
SINCE THE PARTING OF THE RED SEA

You can't cover it up, Pat. It's the biggest story since the parting of the Red Sea!

—REPORTER NED Scott to Air Force Captain Patrick Hendry upon discovering
a crashed UFO embedded in the ice. From the 1951 RKO motion picture,
The Thing from Another World.

When Mack Brazel delivered a box of recovered material to the Roswell sheriff's office, it confounded them as much as it had him. So, Sheriff George Wilcox and Deputy Bernie Clark decided to contact the Roswell Army Air Field. Tuesday, July 8: two days after the military received the Brazel material, which, according to Colonel Thomas DuBose, General Ramey had been "immediately" ordered by his superiors to ship to Washington, D.C.,[1] Blanchard's statement simply said, "The disc was picked up at the rancher's home. It was inspected at the Roswell Army Air Field and subsequently loaned by Major Marcel to higher headquarters." To the media, that made perfect sense: Colonel Blanchard was taking orders from his own boss and superior in the chain of command: Brigadier General Roger M. Ramey, commanding officer of the 8th Air Force at Carswell AAF in Fort Worth, Texas.

It should be pointed out here that for this very purpose of maintaining open communications with the media, the military has public information officers (PIOs). At Fort Worth that person was Major Charles A. Cashon, whose job was to deal with the press. Ramey's own chief of staff, Colonel DuBose, would state that one of the reasons for substituting the weather balloon in the first place was to "get the press off the general's back." DuBose also emphasized that the substituted balloon "couldn't have come from Fort Worth. We didn't launch balloons!"[2] This should cast doubt on Ramey's personal experience with such devices. So the press release would seem more like a preemptive attempt by the general to *personally* make sure the press got it right.

It was just a year earlier that General Ramey had demonstrated his abilities at spin doctoring. He was in charge of the flight crews at the Bikini Atoll Atomic Bomb test ("Operation Crossroads"). Following the dropping of the first bomb there, Ramey informed the press that it "was a complete and unqualified success." In reality, his specially picked crew, commanded by Major Woodrow "Woody" Swancutt (pilot) and Captain William "Wild Bill" Harrison (copilot) in the B-29 *Dave's Dream,* was completely off target and destroyed most of the scientific value of the test.[3] The media accepted the general's assessment. But would they buy his explanation for what was found outside of Roswell a year later?

Consider the following record of events immediately following the announcement of the flying saucer recovery.

- ▶ Within an hour of the original press release, science reporter Dick Pearce of the *San Francisco Examiner* called Blanchard's "higher headquarters" and talked personally with General Ramey. The general told him the object resembled a weather balloon and radar reflector, similar to, Pearce noted, "the ones they sent up every day in Oakland."

- ▶ The *New York Times* claimed that the story began to change "within an hour" of the original press release.

- ▶ Colorado Senator Ed C. Johnson, at least 1½ hours before Ramey's weather balloon story became the official Air Force explanation to the press, was calling the *Denver Post* from Washington and telling them that the object found in New Mexico "... may have been either a radar target or a meteorological balloon."[4]

- ▶ The *Washington Post* reported that Ramey informed the Pentagon press office that "the object was in his office" at that very moment; in fact, he hadn't yet seen it. Shortly afterward, he placed another call to the Pentagon press office and said the object was made of tinfoil and wood and was *25 feet across.*

▶ Ramey then informed a New York newspaper, *P.M.*, that the recovered wreckage looked "like the remains of a target and weather balloon" and was under "high security." Similar quotes appeared in various United Press articles.

▶ Other newspapers quoted General Ramey as saying he "knew it was a weather balloon from the very beginning."

▶ Major Marcel, traveling on orders from General Ramey, arrived in Fort Worth on a special nonscheduled flight from Roswell AAF at approximately 5 p.m. CST with the "flying disc."

Every one of the previously cited interviews with or statements from General Ramey took place three to five hours before the B-29 aircraft carrying Major Marcel and the "flying disc" landed in Fort Worth. Once Major Marcel reported to the general's office, the famous weather-balloon press conference began in earnest.

Once the weather balloon explanation had been given to the press, the second phase of "getting the press off our backs" went into action. As if to punctuate the conclusion of his press conference, General Ramey told another whopper for press consumption that the flight scheduled to take the alleged flying saucer wreckage to Wright Field in Dayton, Ohio, had been cancelled.[5] In truth, it had not been cancelled and arrived at Wright later that day.[6] He then went on a local radio station later that day, not only to reiterate the weather balloon story, but also to suggest that the flying saucer "invasion" that was engrossing the country that summer was related more to the consumption of alcohol than to anything real. Meanwhile, back in Roswell, querying phone calls were already starting to come in from all over the world since the original story that a flying saucer had "landed" near Roswell hit the wire services a few hours earlier. In response, it was decided that the best way to handle the burgeoning press interest was for the principal players to "get out of Dodge," thereby creating a cone of silence about the case. Phone calls for Major Jesse Marcel were answered with a perfunctory "Major Marcel is out of town." That was true, of course, as Marcel had been hustled off to Fort Worth in *Dave's Dream* to be used as an unwitting weather balloon prop in General Ramey's infamous news conference. Callers for the Roswell base commander, Colonel William Blanchard, were told that he was "on leave," which was a lie. In reality, Blanchard had gone out to the crash site north of town to oversee the recovery operation and had also temporarily moved his living quarters into one of the enlisted men's barracks until the excitement had died down.[7] Blanchard also instructed his PIO (Public Information Officer), Lieutenant Walter Haut, who normally would have been the "point person" for dealing with press inquiries, to go home and "hide out," which he did after getting a quick look at the UFO wreckage and dead aliens in Hangar P-3.[8] Completing this runaround circle of silence and misinformation, telephone callers to Lieutenant Haut's office were advised to call the Chaves County sheriff's office, where the deer-in-the-headlights sheriff, George M. Wilcox, could offer only this scripted response, "I can't comment further. I'm working with those fellows over at the base."[9] AP reporter Jason Kellahin, the reporter who interviewed Mack Brazel, also attempted to interview Sheriff Wilcox, whom he knew well because Kellahin himself was from Roswell. But all he was able to get out of Wilcox was that the military had indicated to him it would be best if he didn't say anything.[10] For good measure, to prevent any leakage in case frustrated callers attempted to take matters to higher authorities in the hope of obtaining some information, it was immediately reported in the press that New Mexico Governor, Thomas Mabry, had taken a "sudden vacation to the mountains" with his family.[11] Mabry had been scheduled to meet with Colonel Blanchard the following day (July 9) in an official capacity to sign the Air Force Day Proclamation, but, because both were unavailable, the signing did not take place until the following week—with New Mexico's lieutenant governor, Joseph Montoya, standing in for the governor, who remained unavailable by traveling directly from his "sudden vacation" to a governors' conference in Salt Lake City, Utah.

The final stake into the heart of a now moribund press corps was administered on July 9, 10, and 11 in the form of numerous "balloon photo-ops" all over the country orchestrated by the Air Force and the Signal Corps.[12] Balloon-launch demonstrations designed to impress and convince the press and the American people that these mundane devices (known as "rawin" balloon-borne radar targets, which consisted of one rubber balloon and one six-sided, tinfoil, radar target) were responsible not only for the Roswell Incident but also for the *flying saucer* hysteria sweeping the nation. It worked! The press completely lost interest in the Roswell case for decades, and reports of strange things seen in the sky dropped dramatically. With the press under control, it was now the turn of the eyewitnesses. (See Chapter 11 and Chapter 18.)

11

"SOME THINGS SHOULDN'T BE DISCUSSED, SERGEANT!"

*Well, what they brought back, they had this big old eighteen-wheeler that they brought back from the crash site. Now today, I just laugh at 'em, 'cause they say it was a **weather balloon**—you know, a weather balloon! And I say, "How come they have an eighteen-wheeler out there haulin' a balloon around?"*

—EDWARD HARRISON, Corporal of Security, 1027th AMS,
509th Bomb Group, RAAF (Roswell, N.M., 1947)

Tuesday morning, July 8, 1947, broke sunny and bright very early in southeastern New Mexico, as usual. But there was something going on that was anything but usual to the anxious townsfolk of Roswell. They had been seeing more than their share of strange "things" in the skies overhead, termed *flying saucers* by the press, for the past few weeks, and now there was talk of a crash of one of them north of town. There was also talk of "little bodies" being found.

Patricia Rice of Garland, Texas, owned the Alley Bookstore in Roswell with her husband, who also taught at the New Mexico Military Institute in 1947. She had heard about the crash the day before from her niece, Janet Rice. As she related her remembrances of that time to our investigation: "I can say without much exaggeration that the news of the crash got around that town of 25,000 people in about twenty-five minutes!"[1] Rumors were running rampant at local watering holes such as the Bank Bar in downtown Roswell, where servicemen from the base often went during off-duty hours to mix with other servicemen and with the residents of Roswell. One former

A tractor-trailer was used to transport the inner cabin of the crashed UFO down Main Street in Roswell to the RAAF base south of town, as shown in this computer simulation. (Photo courtesy of John MacNeill and Michael Schratt.)

serviceman who was stationed at Roswell AAF in 1947 said that the town mood at this time was "anxious . . . perhaps *scared* would better describe it."[2] And so far, the Roswell air base had said nothing, and the local media—the *Morning Dispatch* and *Roswell Daily Record* newspapers, as well as KGFL and KSWS, the local radio stations—had said the same. Unnoticed that morning, except by a few RAAF airmen who had gone to the base chow hall for the 5 a.m. early-bird breakfast, was the eighteen-wheel "lowboy" parked outside. Staff Sergeant George D. Houck of the 603rd Air Engineering Squadron had checked it out of Squadron T (the base motor pool) a few minutes earlier and, before heading north, stopped at the chow hall for the early-bird. Similar to many other married RAAF airmen at the time, Houck was living in temporary base housing at the former WWII German prisoner-of-war camp located at Orchard Park, just south of Roswell. Houck ate alone that morning and, after finishing, put his tray in the "clipper" and quickly left.

At midmorning, with the morning staff meeting over, Colonel William Blanchard, commanding officer of not only the 509th Bomb Group, but also the Roswell air base itself, summoned the base PIO, First Lieutenant Walter Haut, into his office and told him to issue a press release for dissemination to the local media. The release, which began, "The many rumors regarding the flying disc became a reality yesterday when the intelligence office of the 509th Bomb Group of the Eighth Air Force, Roswell Army Air Field, was fortunate enough to gain possession of a disc . . ." was then hand-delivered by Haut to the four Roswell media outlets around noontime.[3] But it would be another three hours before the shot-heard-'round-the-world headline, "RAAF Captures Flying Saucer on Ranch in Roswell Region," hit the newsstands.

It was too late in the day for the *Roswell Morning Dispatch,* which would have to wait until the following day, July 9, to weigh in on the burgeoning Roswell "crashed saucer" story. Instead, by default, it would be the afternoon newspaper, the *Roswell Daily Record,* that would make history that day. Richard Talbert, a *Roswell Daily Record* paperboy in the summer of 1947, recalled for us a day in early July 1947 when he had just picked up his batch of the *Record.* It was somewhere between 3 p.m. and 3:30 p.m. as near as Talbert could recall. He was plying his trade in the vicinity of the *Roswell Daily Record* building at 4th and Main streets in downtown Roswell along with a number of other paperboys, when he looked up and saw something he had never seen before in his young life. Heading south down Main Street was a military convoy composed of one large, eighteen-wheel, lowboy or flatbed trailer protected by an escort of jeeps in front of and behind it, each carrying a contingent of armed MPs. But it was the trailer—or what was on it—that really caught Talbert's attention that day. "The lowboy had a tarp on it, and there was something under the tarp. Whatever was under there appeared to be oval-shaped. I don't recall now how I did it, but I was able to get a quick look under the tarp. I think it must not have been securely tied down on one end, or it just came loose, and it flapped up briefly as it went past me. Anyway, I saw a silver, oval-shaped *something* that was approximately 4 to 5 feet wide by about 12 feet long and 5 to 7 feet high. It had a dome on it, but it was damaged because it was cut off at one end."[4] Bob Rich was also a paperboy that summer and saw the convoy pass "right through the center of town" that day.[5]

Paul McFerrin was a Roswell preteen in 1947 who was out with friends, Floyd and Lloyd Carter and Charlie Webb, that afternoon. "This was right

Dennis Chavez, New Mexico's junior senator, flew to Roswell to warn the Anaya brothers to remain silent "or else!" (Photo courtesy of the U.S. Senate Historical Office.)

at the time of the flying saucer incident," McFerrin said, "about a week or so after a very severe lightning storm that lasted two days. We were walking down Main Street when we saw this big, military flatbed transporting an egg-shaped object through town, obviously heading for the base. The flatbed trailer had a tarp over the object but you could pretty much tell what shape the object underneath was. It was escorted by MPs in jeeps who were holding machine guns. Everybody in town knew about the crash."[6] A few miles farther south, Jobie MacPherson was in the middle of completing a roofing job for his employer, Lynn Everman Construction Co., when he spotted the convoy. "It was coming from the north heading toward the base and went right past me. Jeeps and a flatbed truck. I could see mangled metal sticking out on the flatbed and something else that had a conical shape to it, like a pod or something."[7]

In 1947, Private First Class (PFC) Rolland Menagh was a twenty-year-old MP in the 390th Air Service Squadron (ASS) of the 509th Bomb Group stationed at the Roswell Army Air Field. Unlike the MPs of the 1395th Military Police Company at the RAAF, who were used for general base security under the command of base Provost Marshal Major Edwin Easley, the MPs of the 390th ASS, under the command of Major Robert Darden, held higher security clearances, which were required for guarding the Silverplates—the atomic bomb–carrying B-29s. When the UFO impact site was discovered north of town by civilians on the morning of July 7, MPs of both the 390th and 1395th were rushed to the scene to secure it. MPs of the 1395th were posted along the western edge of Highway 285 from Roswell in the south all the way to the hamlet of Ramon in the north in order to prevent civilians from reaching the crash site five miles west of the highway. The MPs of the 1395th also composed the outer ring of armed guards circling the crash site. MPs of the 390th formed an inner ring of armed guards circling the crashed UFO and its occupants. Because they would have been close enough to see the crash scene up close and personal, the higher security clearances possessed by the MPs of the 390th were required for this task.

Menagh would later go on to become a security specialist for the Air Force Office of Special Investigations (AFOSI) after the Air Force became a separate branch of the armed forces. He passed away in 1996 at the age of seventy, but our investigation located and interviewed Rolland Menagh Jr. in 2005. He confirmed his father's presence at Roswell AAF in 1947 as well as his participation in the UFO recovery operation north of Roswell. "My father first told us about it in the 1960s," he said. "He was an MP who guarded the UFO crash site north of Roswell. He saw the ship, which he described as being round or egg-shaped and seamless." Menagh Jr. said that his father did not see any bodies, but his brother Michael thought their father had told them there were *three* dead bodies:

> *He said that the spaceship was loaded onto an eighteen-wheeler with a tarp covering it and then driven right through the center of town down to the air base. My father said that he had accompanied it in a jeep all the way from the crash site to a hangar at the base where it was deposited. Afterwards, he was sworn to secrecy, and when he left the Air Force, he was reminded about the episode and told to "Keep quiet, or else!" Later, in retirement, he periodically received visits from military types in dark suits over the years who were obviously keeping tabs on him.[8]*

The late John Tilley was a retired Air Force master sergeant who was interviewed in his Roswell, New Mexico, home by Tom Carey in July 2008. Tilley, who did not arrive at the RAAF until a few months after the July 1947 events as an eighteen-year-old, nevertheless heard the stories and rumors still swirling around the base at the time about the crash. One rumor even had the one "live alien" escaping from the base only to be shot dead by panicky MPs at the main gate when it suddenly reappeared, after startling residents of a nearby trailer park. In 2007, Tilley self-published a book about his knowledge of the 1947 Roswell events.[9] In it, Tilley tells of a brother-in-law of his, James W. Storm, who, as a retired Army sergeant a few years ago, told Tilley that he was involved in the transport of the wrecked UFO from the crash site back to the air base. In July 1947, Storm was a member of the RAAF base fire department when the call came in to head out to a location north of town. According to Storm, the group of vehicles involved was made up not only of military firemen but City of Roswell firemen as well. The fact that Roswell firemen were called out suggests that the location of the site the group was heading for was a site located in Chaves County—not in Lincoln County where the Brazel/Foster Ranch "debris field site" and the Dee Proctor "body site" were located.

As the group of vehicles left Highway 285 and headed west to reach the site, they were instructed to park off-road out of sight of passersby and await further orders. After a few minutes, a "snub-nosed tractor and lowboy flat trailer showed up." On the back of the lowboy was a tarp that was covering something. According to Storm, in his own words to Tilley, it was "a saucer part so big (that) it was covered." As the lowboy went by toward Highway 285, the firemen were ordered to accompany the lowboy back to the base. Storm then told Tilley of "escorting a large piece of the saucer along Main Street through Roswell . . . in the late afternoon!" (NOTE: Tilley refers in his book to Storm's "large piece of the saucer" as the stricken craft's "cockpit," while we have referred to it as an "escape pod" or the "inner cabin.") Putting an exclamation point on his story and perhaps providing some insight into the importance of what was taking place so many years before, Storm informed Tilley that the assistant fire chief had gone along on the trip, and that the assistant fire chief had been told, "If he 'cooperated' . . . he would be made fire chief. Which he was!"

A crew from the RAAF fire station [shown here], to the exclusion of one from the City of Roswell, was sent to the UFO impact site just north of town to handle the situation.(Photo courtesy of Tom Carey.)

Highway 285 runs north and south right through the center of

Roswell, where it becomes Main Street, straight down to the old air base at the south end of town. Another relative of John Tilley's, a retired MP by the name of Bill Blair who was at the base in 1947, confirmed that the ultimate destination of the lowboy convoy was Hangar P-3 (Building 84 today) at the east end of the base tarmac. According to Blair, Hangar P-3 was considered a "security building" when the need arose, and, "anything too large for a safe went there."[10] We know from other witnesses cited herein that Hangar P-3 (a.k.a. the "Big Hangar") was the central focus and repository of the UFO recovery operation.

The testimony of Chester P. Barton, who passed away in 2000, illustrates the various levels of security present on the Roswell base in 1947 and the precise distinction between the 1395th MP Company and the 390th Air Service Squadron regarding security matters. Barton was a first lieutenant in the 1395th serving under Captain Beverly H. Tripp when he was ordered to report to his unit's commanding officer, the base provost marshal, Major Edwin Easley. Subject: Recovery Operation North of Roswell. Soon thereafter, Barton and Tripp were motoring in a jeep carryall to an undisclosed flat, open desert area about forty-five minutes north of town. Barton had no idea what to expect, but the armed checkpoints and rumors about a downed spaceship served to fuel his curiosity. But curiosity would not account for the armed security roadblocks Barton ran into upon arriving at the heavily guarded scene before him. Though he was an officer, Barton was allowed no closer than 150 feet from the wreck. He would perform no duties nor engage anyone with a "need to know" what was going on. Barton was restricted to the cheap seats with the sole mission of reporting back to Easley what he saw and what his impressions were. No doubt Easley was very pleased, because his lieutenant, Barton, with no prior briefing, came away with the conclusion that it was merely the wreckage of a B-29 bomber.[11] Personnel on the periphery without the highest of security clearances would be less inclined to believe the rumors. Better to have them think it was just "one of ours." Still, Barton was never able to identify any of the metallic debris that he did see. But, more importantly, he did confirm an actual crash site not too far north of Roswell, an intact craft of some sort, scorched ground, and men testing the site for radioactivity. Because records show that there had been no crash of any military or civilian aircraft north of Roswell during that time frame, Barton's observations deal a fatal blow to the current Air Force explanation that rubber balloons and tinfoil radar targets from Project Mogul were responsible for the Roswell Incident. Armed security details and radioactivity concerns aside, Barton stated that he had witnessed *metallic* debris at the crash site—he thought it might have been B-29 wreckage, but that was just a supposition given his distance from the craft itself—and burn marks on the ground, both of which rule out any type of "balloon crash," be it weather, Mogul, or otherwise.

Earl V. Fulford was a twenty-one-year-old in July of 1947, a staff sergeant with a top secret clearance in the 603rd Air Engineering Squadron (AES) at the RAAF. He had arrived at Roswell in 1946 upon reenlisting in the Army Air Corps after spending WWII in the U.S. Navy. He worked as an aircraft mechanic and forklift operator in the hangar located next to Hangar P-3, which was the center of the UFO crash-retrieval operation. One of the main tasks of the 603rd was the overhauling of the B-29's temperamental Wright Cyclone engines. Because each engine of the four-engine Boeing B-29 Superfortress bomber had to be overhauled every sixty-five hours due to its tendency to over-heat and catch fire, it was a constant activity for the men of the 603rd. Fulford had seen his friend George Houck at the 5 a.m. early-bird breakfast, but didn't get a chance to talk to him, and he didn't think too much about it when he saw Houck drive off in the lowboy. "I just thought that George was being sent to pick up a wreck somewhere," he said. "Nothing special."

Things would soon change, however. A number of the civilian mechanics who worked in the same hangar as Earl Fulford began peppering him with questions about what was going on as soon as he arrived for his shift that morning. It was, in fact, how Fulford first learned of the crash. "They kept asking me all day long, 'C'mon, tell us. What crashed? You must know. The word in town is that it was a spaceship with bodies of little spacemen.' I didn't know what to tell them other than I didn't know anything about it, but it sure got me to rethinking about George Houck and where he might have been going with the lowboy after breakfast."[12] Staff Sergeant Harvie L. Davis, also of the 603rd AES, remembered hearing the "scuttlebutt" at that time about the crash of a flying saucer. "Stories were going around, and I didn't—and still don't—doubt the people involved. I believe that it was a UFO."[13] John Bunch was also a member of the 603rd AES. "Everything was hush-hush," he said. "We all knew that something was going on, but we didn't know what. A lot of planes were coming

Earl Fulford [L] with coauthor Tom Carey on the former J. B. Foster Ranch, which Fulford hepled clean up in 1947. Fulford passed away one month after this photo was taken. (Photo courtesy of Tom Carey.)

in and going out, and the airstrip was shut down for a period. The base went into lockdown, and they checked us real close going in and out."[14] Eugene C. Helnes was a PFC in the 603rd AES in July of 1947. "It was definitely not a balloon," he told us. "I knew fellows who were out there at the site to clean it up. All the talk was of a crashed saucer—right up to the time that I left the base in mid-1949."[15] Reflecting today on his then-growing awareness that something highly unusual had taken place, Earl Fulford told our investigation in 2006, "Master Sergeant Hardy [First Sergeant Leonard J. Hardy of the 603rd AES] had a look on his face like he knew something that he wasn't telling us. It was clear that scuttlebutt was out in town almost immediately, especially about the bodies, but on base no one was telling us anything."

With the completion of his shift at 4 p.m. on this afternoon of July 8, Earl Fulford and a few friends left the hangar with the intention of heading over to the NCO Club across the base for a couple of cool ones to help them deal with the rumors and contemplate their futures. As they approached Fulford's car, parked in the parking lot across the street from the hangar, he could see a large tractor-trailer rig coming down the street in his direction, apparently heading for Hangar P-3. "It had come through the main gate, past the parade ground, and made a left turn toward the hangars," he said. When the rig got close enough, Fulford could see that it was pulling a lowboy trailer, and that the lowboy was carrying something under a tarp that was "about the size and shape of a Volkswagen Beetle." Fulford now also recognized the driver of the rig, none other than his friend, George Houck. According to Fulford, one of Houck's base duties was that of a "reclamation specialist," meaning that whenever there was a wreck somewhere to be hauled back to the base for disposal—an aircraft, a staff car or truck, and so on—the job was his. Seeing Houck behind the wheel of the rig, Fulford waved his arms for Houck to stop. With the engine idling, Houck leaned out of the driver's window slightly to see what Fulford wanted.

EF: Where you been, George?

GH: Up north.

EF: What'cha got under the tarp?

GH: I can't tell you.

EF: Why not?

GH: They told me not to say anything.

EF: Let's have a look anyway.

[Having said that, Fulford and his friends headed to the back of the lowboy and started to lift the tarp; seeing what was up, Houck quickly jumped out of the cab and ran after them.]

GH: No! No! No! You can't do that! I'm not allowed to look under there, and no one else is allowed either.

EF: Why not?

GH: I don't know, but I've been sworn. That's all I know.

With that, Houck jumped back in the cab, put the rig in gear, and drove into Hangar P-3.[16] Recalling his run-in with the lowboy these many years later, Earl Fulford had no doubt that it occurred just the way he described to us, and that the driver was his old 603rd friend George Houck. Fulford then related a little story to us involving Houck and himself that had taken place a few months before their July 1947 "close encounter." Another fellow 603rd airman by the name of Eldo Heath had apparently gone AWOL by "appropriating" a private airplane from the Roswell Municipal Airport, where he had been taking flying lessons, and flying it all the way to the state of Georgia. It was several months before Heath was located, and Fulford and Houck were summoned by their commanding officer to travel to Georgia by transcontinental train and return Heath to Roswell, which they did, with a stopover at Sportsman's Park in St. Louis to take in a Cardinals game. The story had a happy ending—for Heath, at least—in that the episode did not permanently stain his military record, as he would later go on to become a pilot in the new Air Force.

Our investigation located George Houck in 2005, who was ninety years old (now deceased). He claimed not to remember his former 603rd buddy, Earl Fulford. But after reprising him with the saga of Eldo Heath, Houck said that he now remembered Earl Fulford. If you interview enough people, you can tell pretty early in the interview of how it's going to go. This one was one of those that was not going to go well, and Houck didn't disappoint. Perhaps sensing where we were heading with our questions, Houck became evasive. He claimed not to remember at all the incident involving Fulford and the tarp-covered object on the lowboy he was allegedly hauling to the big hangar in July 1947. "I don't remember that at all. It was a long time ago."[17] That was as far as we could get with George Houck, so we then asked Earl Fulford if he wouldn't mind giving Houck a call to relive old times and perhaps obtain some of the answers we were not able obtain. Fulford agreed. After his initial phone call to George Houck, Fulford gave us the results:

> Houck first claimed not to know me. Then, after a while, he finally relented, and we started to talk. Then I asked him if he remembered the incident in front of the hangar with the lowboy, and there was a long silence. Then Houck said, "What do you want of me?" I told him that I just wanted to know if he remembered it. After another long pause, Houck then shocked me a bit with, "Some things should not be discussed, Sergeant." Then he hung up.

Fulford did not give up, however, and would call George Houck another half-dozen or so times over the course of the next two years. Houck, however, has retreated to the final fallback position of a defendant not wishing to incriminate oneself, but not wanting to "take the fifth" either, by simply telling Earl Fulford, "I can't remember."[18]

That was not the end of Earl Fulford's brush with history in the Roswell story, however. We have determined, from interviewing surviving family members and friends of Mack Brazel, that he had been sequestered at the

RAAF for about a week (July 8 to July 15), while the cleanup operation was taking place at his Corona ranch. The sea of strange debris from the crippled UFO that covered his sheep pasture had to be cleared, and as near as we can figure, this operation began on July 7 and was completed by July 15. It was the late Robert E. Smith who told our investigation in 1990 how this was accomplished. Smith, who was a sergeant in the First Air Transport Unit (the "Green Hornets"), said that several groups of sixty or more men were trucked out to the ranch, where they were told to "pick up everything not nailed down" in the pasture and place it in wheelbarrows for collection at a central point. The men went back and forth over the pasture until all of the larger pieces of debris were picked up. Then another group of sixty or so men would come in and go over the same ground, picking up what the previous group had missed. That's how it was done. Smith also said "All of a sudden, there were a lot of people in dark blue suits on the base that I did not recognize." Crews for this purpose were brought in from Fort Bliss in El Paso, Texas, and possibly from White Sands near Alamogordo, New Mexico. Fulford was also enlisted.

The early morning of Wednesday, July 9 found Earl Fulford enjoying his 5 a.m. early-bird breakfast at the chow hall, as usual. Not so usual was the sudden appearance of Master Sergeant Earl Rosenberger, heading for Fulford's table. Rosenberger was a member of the 603rd AES, an old-time sergeant whose date of service went all the way back to World War I. Looking right at Fulford, Rosenberger growled, "I need a detail. Let's go!" It was too late for Fulford to look away and pretend that he didn't hear him (an old ploy that all potential "volunteers" in the military employ to lessen their chance of being chosen), plus Sergeant Rosenberger was looking right at him. Fulford reluctantly left his early-bird "SOS" (military jargon for a type of breakfast dish) for the maw of the clipper and followed Rosenberger out the door where he joined fifteen to twenty other enlisted men and noncommissioned officers waiting outside. A military bus then pulled up, and they were all told to get on. As Fulford told us:

> The bus ride took about two hours. The site was northwest of Ros-
> well. We went north up Highway 285, then west on the Corona
> road, which was a gravel road back then, past a little schoolhouse
> and some other structures [this was in all likelihood the present-day
> ghost town of Lon, which is no longer on the map], and then turned
> south onto a dirt road. I remember seeing a little house [the Hines

House], which was not far from where we were going. When the bus stopped, we were told to get out. A major was in charge [this was Major Edwin Easley, the RAAF provost marshal], and armed MPs ringed the site, which was situated at the base of gently sloping hills. Sergeant Rosenberger then handed each of us a burlap bag and told us to "police up" the site and put anything that we found in the bags.

Fulford could only recall that the site to be "policed" encompassed "hundreds of yards." Fulford and crew had arrived at the Brazel debris field site on the J. B. Foster ranch. This was not the same site (known as the "impact site") from which the egg-shaped vehicle or "pod" that was transported through the center of Roswell on the lowboy was retrieved. That site was much closer to Roswell. This site, known simply as the "debris field," was nearer to Corona in Lincoln County and was the location over which the outer surface of the unknown craft came apart for reasons unknown, raining debris down on the desert floor below. What remained stayed in the air a few more seconds, ultimately falling to Earth at the impact site.

It was evident to Fulford and the rest of his crew that theirs was not the first crew to police the site. They could see tire tracks from heavy trucks all over the landscape, and the amount of recoverable material appeared to be sparse. Clearly, a full-scale recovery operation had commenced sometime before. One of the drivers of those heavy trucks was PFC Frank M. Martinez of "T" Squadron at the RAAF. This was the base motor pool that provided the cars, jeeps, trucks, and other equipment for such an event. According to Martinez, as ordered, he would drive an empty army truck out to the Foster ranch, and then jump into the cab of a loaded vehicle and transport the secret cargo back to the base. Once there, he would drop it off at Hangar P-3. Then, he would head back to the ranch with another empty truck ready to receive more crates for immediate transit back to Roswell. All of this was verified by Martinez's wife, Mary, who emphasized, "My husband drove the truck hauling the 'flying saucer' material for *two full days!*"[19] They were never once told what it was that they were supposed to look for or what had happened there—just pick up anything that was not part of the landscape. "We knew from the day before that something had crashed up there, so we figured that this must have been the crash site," he said. The men formed a single line abreast with a few feet between each man and began traversing the pasture, picking up anything that was "not natural" and placing it in the burlap bags. They performed this procedure over and over again in every conceivable direction on the site until 4 p.m., after which each man was told to empty his bag into a wheelbarrow near the entrance to the pasture. When this was accomplished, the men boarded the bus and returned to Roswell, never to return to Brazel's pasture again.

Recalling the experience for us, Earl Fulford remembered that during the entire policing operation, he picked up only seven pieces of debris. But they were enough to confirm his original suspicions regarding the site. He told us:

> I picked up small, silvery pieces of metallic debris, the largest of which was triangular in shape, about 3 to 4 inches wide by about 12 to 15 inches long. It looked like thin, light, aluminum foil that flexed slightly when I picked it up, but once in the palm of your hand, you could wad it up into a small ball. Then, when you let it go, it would immediately assume its original shape in a second or two—just like that! That was the only type of debris I saw that day. I thought to myself, "Hey, this is neat. I'm going to keep a piece for myself." But they searched us thoroughly when we got back to make damned sure that none of us had anything. Nobody picked up anything of size. We didn't see any other type of debris or pieces of debris with writing on them, and we didn't see any bodies. We also did not see any balloons or balloon material. They launched weather balloons from in between barracks where I lived back on the base every day. I was familiar with them, and the debris wasn't from one of those. When we got back to the base, everything that we picked up was taken to Hangar 3. We were then lined up and told one by one by our first sergeant [Master Sergeant Leonard Hardy] in no uncertain terms that we didn't see anything, and we didn't say anything; and if we did from that point forward, we might be courtmartialed. A few days later, I think it was on Saturday, our entire squadron was called together for a special meeting in Hangar 1 where we were addressed by our squadron commander, Major Harry Shilling. Also present was our second-in-command, Captain Earl Casey, and a glowering First Sergeant Hardy who had been in my face a few days earlier. Captain Casey gave a cautionary admonition to everyone present not to talk about anything they might have seen or heard in the past few days, but Major Shilling got right to the point: "You didn't see or hear anything. Nothing happened!"[20]

A few days earlier, PFC Harry Girard of the First Air Transport Unit (ATU) had found himself in a similar meeting in another hangar on the base about to be addressed by his commanding officer, Lieutenant Colonel James R. Wiley. Girard had no idea why he was there or what the meeting was about. Colonel Wiley addressed the group: "If you know something, you keep your mouth shut! If you don't, you may find yourself at Leavenworth (Prison), and you can read all about it there." Colonel Wiley never told the group what that something was, or why talking about it could land them in Leavenworth. PFC Girard and perhaps more than a few others that day had no idea what their CO was talking about. Word probably had not yet reached them about the multiple flights made by aircraft of the First ATU that had flown UFO wreckage and "little bodies" recovered from the crash site out of Roswell in the preceding days. Interviewed in 2005, Harry Girard said that the meeting in the hangar "was unusual, because I had no idea what he was talking—actually threatening us—about. I do now, and maybe in another fifty years they will tell us what happened."[21]

Threats seemed to be the order of the day, the modus operandi for handling the problem. Lieutenant Steve Whalen was a twenty-eight-year-old navigator in one of the bomb squadrons of the 509th Bomb Group at the RAAF during that second week of July 1947. Although nothing official had been said to base personnel, Whalen and his friends knew that something was going on at the big hangar, but they were afraid to inquire about it. "There were a lot of people who were real scared who were not talking," Whalen's son told us. "There were also a lot of planes taking off and landing in the middle of the night." A special meeting was then called by his bomb squadron's commanding officer—not to explain what was going on, but to issue a warning: "Stay away from the big hangar area no matter what, or you will run the risk of being shot on sight!" That was all. Until that meeting, Lieutenant Whalen had thought that the base commotion might have had something to do with

"Goddard stuff" (Dr. Robert Goddard, the Father of American Rocketry, had conducted his rocket experiments just outside of Roswell). Whalen died in 1998, and he apparently followed his CO's orders to the letter.[22]

By Monday morning, July 14, 1947, things had pretty much calmed down at the Bank Bar in Roswell, as well as at Roswell Army Air Field. The press and the public had quickly lost interest in the Roswell crashed "saucer" story a few days earlier when General Roger Ramey issued his famous press release. In addition to the airmen on the base, the ranchers in the Corona area and key civilians in Roswell had been threatened into silence. The crash sites had been cleaned up, and most of the wreckage, including the bodies, had been shipped out. Life had returned to normal. Or had it?

At 2 a.m., the sleeping Sergeant Earl Fulford was suddenly awakened to the sight of a flashlight shining in his eyes and the unsmiling face of Master Sergeant Larry Sanchez staring down at him. Fulford's first thought was, "What's this? Did I miss early-bird?" Sergeant Sanchez barked, "Get dressed and follow me!" Still in that semi-shock state from being surprised in the middle of the night, Fulford dressed as quickly as he could and followed Sergeant Sanchez out of the barracks into the mist of the early morning. Sanchez led Fulford to the front of Hangar P-3 where a large wooden crate was sitting all by itself. Also sitting in front of the hangar on the tarmac was an idling C-54 aircraft with its cargo door open. In a voice loud enough to be heard over the idling engines, Sanchez ordered Fulford, who had recently taken over the duties of a forklift operator from someone else who had been transferred, to load the wooden crate into the C-54. After locating and firing up the fork-lift, Fulford carefully lifted the crate onto his machine, drove it over to the waiting aircraft, and deposited it inside the yawning cargo hold. With that, the cargo door slammed shut, and Fulford was told to leave as the plane taxied away toward the flight line. "I had no idea what was going on. No one told me a thing. I just followed orders. I do know one thing, however. Given the size of the crate, which was about 7 feet by 7 feet by 7 feet, whatever was in there had to weigh almost nothing. Either that, or they woke me up in the middle of the night to load an empty crate. I could tell that right away when I first lifted it."[23] After returning and shutting down his forklift, Fulford left and walked back in the direction of the chow hall. It was almost time for early-bird.

In July 2008, while we were in Roswell for the yearly UFO festival, we introduced Earl Fulford to Bill Ennis. In 1947, Sergeant William C. Ennis was a flight engineer with the 393rd Bomb Squadron at the RAAF. Hangar P-3 was the primary facility of the 393rd at the time of the incident, as well as the primary receiving facility for the wreckage and bodies recovered from the crash. It stood to reason that anyone who worked in that hangar might know something. We had interviewed Ennis years before in person in 1992, but all he would offer was to laugh everything off as just a weather balloon. After years of denial, and after meeting Earl Fulford, Ennis allowed, "Whatever Earl told you is all true." It would take an additional meeting later in 2008 with Schmitt to obtain Ennis's final confession: "It was a spaceship. After all these years, I still don't know how that ship flew. *There was no engine!* Before I go, I'd like to know."[24]Ennis passed away in 2010 before we *knew.*

So, what became of the "escape pod" or "inner cabin" from the UFO that was seen on the lowboy that Sergeant George Houck deposited in the "big hangar"? When last seen, it was still in the hangar. After that . . . ? We have the accounts from the men who boxed up pieces of the wreckage and the bodies for shipment from the RAAF, but nothing that sounded like a ship. We also had the testimony from a few who claimed to have stood guard at Hangar P-3 months after the July incident, but we didn't know what to make of these puzzling statements. In 1999, we also interviewed a fellow by the name of George Newling who claimed to have seen a "teardrop-shaped" craft in the belly of a B-29 that he was checking out prior to takeoff from the RAAF in November of 1947. We didn't know what to make of Newling's story at the time, because his time frame seemed to be wrong, because we had always assumed that all of the crash wreckage had been shipped from the RAAF at the time of the recovery. We also didn't yet have the accounts of

Earl Fulford regarding George Houck, or of those who had witnessed the lowboy transporting an "egg-shaped craft under a tarp" down Main Street in Roswell. Once we had these witness testimonies, though, we had our answer.

PFC George Newling had arrived at Roswell Air Field (the interim designation of Roswell Army Air Field after September 18, 1947, until it became Walker Air Force Base on January 1, 1948) in the fall of 1947, just a few months after the Roswell event. In three interviews with us, Newling related that he had been a flight mechanic at the RAAF. He recalled how, in November 1947, he had been going through his regular procedures for "pre-flighting" his aircraft, a B-29 Silverplate (number #44-27304) nicknamed "Up an' Atom," prior to takeoff. But this time would prove to be different for the unsuspecting private. It was already different in that Newling noticed security guards patrolling around his aircraft. Although this was a restricted area, it was unusual to see this level of security attached to a single aircraft. While inside the belly of the B-29, Newling was about to check out the aircraft's two bomb bays when he saw *something* he had never seen before and would never see again! "It was shaped like a 'teardrop' and slightly damaged, but still intact. It appeared to be metallic and grayish in color, about 4 to 6 feet high and 10 to 12 feet long, and it had what appeared to be small, hexagonal cells or plates

running the length of it on what I assumed was its underside. I didn't know what this thing was. I still don't. Never saw anything like it before—or since, for that matter." Newling's stupefaction was abruptly terminated by a booming, "Who are you, and what are you doing here?" One of the armed security guards was now inside the plane, glowering right in Newling's face! Newling quickly showed the guard his credentials that identified and entitled him to be where he was, but apparently it wasn't enough for the situation. "You're not supposed to be in here!" With that, Newling was physically removed from the B-29 and marched off to the provost

Former RAF PFC George Newling going over the 1947 RAAF Base Yearbook at the International UFO Museum and Research Center in Roswell in a 2008 photo. (Photo courtesy of Tom Carey.)

marshal's office. Accepting Newling's explanation for his presence
in and around the "Up an' Atom"
that day, Major Edwin Easley
ordered Newling to report directly
to his squadron headquarters and
not return to the flight line until
permission orders were specifically given. It would be three days
before Newling was allowed to
return to the flight line to resume
his duties.[25]

The B-29 dubbed "Up an' Atom" flew the pod-
like "innercabin" out of Roswell a few months
after the UFO incident. (Photo courtesy of
Steve Pace/The Crowood Press.)

We do not know the destination of the B-29 "Up an' Atom"
that day but assume it to have been Wright Air Field in Dayton, Ohio.
Other locations that we have heard throughout the years from people who
claimed to have seen the "Roswell UFO" include White Sands, New Mexico,
Muroc (Edwards) Air Force Base in California, CIA headquarters in Langley, Virginia, and, of course, the infamous "Area 51" at Nellis Air Force Base
in Nevada. We don't know for sure, but what we now know is that the intact
portion of the "Roswell UFO" was kept in Hangar P-3 at the Roswell base
for several months before it was finally flown out in the forward bomb bay of
a B-29 with a name on its nose of "Up an' Atom."

POSTSCRIPT

Sadly, since the publication of the first edition of this book, former Sergeant
Earl V. Fulford passed away on August 3, 2008, at the age of eighty-two. A
month earlier, Earl got to fulfill his late-in-life dream of returning to the
scene of it all, Roswell, New Mexico. He got to tour the old base, pointing
out for us where he worked and where things used to be. He also took us to
the exact spot where sixty-one years earlier he had hailed George Houck's
tractor-trailer "lowboy" with its "strange cargo" hidden under a tarp. Earl was
also taken to the Foster/Brazel ranch debris field site and confirmed for us
that it was indeed the location where he was ordered along with others to

"police the area" of anything "not natural and not nailed down." While in Roswell, Earl gave a presentation of his remembrances to a Saturday night standing ovation at the UFO Museum, answered questions, signed copies for purchasers of our book, appeared on CNN's *Larry King Live!* and participated in the filming of a segment about Roswell for the History Channel's *UFO Hunters* show. Upon returning home to Florida from California, after filming a follow-up segment for the *UFO Hunters,* Earl became ill and passed away in-hospital a week later. According to his wife, Mary, he passed away in his sleep and did not suffer. And the last night of his life, he was doing something he loved—watching NASCAR on TV. His ashes are interred at Bushnell National Military Cemetery in Florida.

Since Earl Fulford's passing, we called his former friend, George Houck, several times, first to inform him of his old friend's death, and second to see if he would at last confirm as true Earl Fulford's account of his participation in the Roswell Incident. In our discussions with Geoge Houck, he changed his responses to our questions from "I handled a lot of wrecks back then," to "I don't recall anything like that," to "I don't remember anything."

QUESTION: "Do you remember anything *at all* about an alleged crash of a UFO near the town of Roswell in 1947?"

ANSWER: "No."

George Houck was a thirty-year Air Force veteran and retiree living on a government pension when we asked him that question. Just before his passing, Houck's daughter promised to try to get answers from her father to the questions that we had been asking him. We never heard from her again.

Earl Fulford's final legacy is that he stood tall when it counted, for the sake of history and for truth. He spoke slowly with a down-home folksiness, yet with a clarity and conviction that held your attention until he was finished. His untimely passing reinforces our awareness of the ever-shrinking pool of firsthand eyewitnesses to one of the most significant events of all time. How many are still left? We don't know for sure, but time is not on our side.

R. I. P. Earl Fulford, February 2, 1926-August 3, 2008.

12

LOANED BY MAJOR MARCEL
TO HIGHER HEADQUARTERS:
FROM COMPLICITY TO COVER-UP

hortly after 1 p.m. on the afternoon of July 8, 1947, a silver Boeing B-29 Superfortress bomber nicknamed *Dave's Dream* taxied up to the flight operations building at Roswell Army Air Field. This was no ordinary flight, even by RAAF standards.

It had been personally ordered by none other than Colonel William Blanchard, the base commander. Its command crew was not one of the regular ones chosen from three constituent bomb squadrons on the base: the 393rd, the 715th, or the 830th.

Instead of the usual cast of young lieutenants and captains, this flight was to be commanded by lieutenant colonels, including the Roswell deputy base commander himself (just below Blanchard in the RAAF command structure), two majors, and a captain—all from Blanchard's close staff. The enlistees on board were all experienced NCOs—tech sergeants and master sergeants—not the usual mixture of privates, corporals, staff sergeants, and others who were included as a matter of course in normal crews.

The flight was booked to go all the way to Wright Field near Dayton, Ohio (Wright Field would later that September combine with the adjoining Patterson Airport, home of the Foreign Technology Division, to form Wright-Patterson Air Force Base), after a "preliminary" stop in Texas at the Fort Worth Army Air Field, headquarters to the Eighth Air Force under the command of Brigadier General Roger Ramey, under whom Blanchard and the 509th directly served.

This flight was to have a special guest on board, Major Jesse Marcel, the head of intelligence for the 509th at Roswell. Marcel was carrying something special to higher headquarters on Blanchard's orders and was told before departure that the flight was to go directly to Wright Field. Marcel had earlier that morning briefed all in attendance at a special staff meeting after his return from a two-day fact-finding trip to the high desert of Lincoln County, where he had encountered a sheep pasture full of strange, shattered, foil-like, plastic-like debris. He brought as much of it back to Roswell with him as he could fit into his baby-blue 1942 Buick Roadmaster convertible.

The B-29 dubbed *Dave's Dream* on the RAAF tarmac transported Major Marcel and some of the wreckage that he and Captain Cavitt had recovered to Fort Worth AAF on July 8, 1947. (Photo courtesy of the 1947 RAAF Yearbook.)

It would be soon after taking off from Roswell that Marcel was informed about a short layover at Fort Worth, where, as we know, soon after the one-hour flight had touched down, General Ramey announced to the world that a flying saucer had *not* been recovered by Marcel and the 509th command at Roswell, but merely the misidentified remains of a very common rubber weather balloon and kite-like, tinfoil radar target.

To seal the verdict, several pictures were taken of both Marcel and Ramey (by himself and with his chief of staff, Colonel Thomas DuBose), each posing with these mundane items on the carpeted floor of the general's office.

To gild the lily, base weather officer Irving Newton was personally called and ordered by Ramey to report to his office and identify the remains as a weather balloon and a rawin-type radar target and pose for a picture as well. A number of these photos went out over the wire services and were picked up by many newspapers across the country as a final solution to the previous day's excitement. The loud sucking sound then heard in and around Fort Worth was the air going out of a big news story as General Ramey "emptied" Roswell's saucer.

The July 8 flight was actually the second confirmed to transport debris out of Roswell since the crash. The first flight took place two days earlier, on July 6. According to DuBose, it was ordered by General Clements McMullen, deputy commander of Strategic Air Command (SAC) at the Pentagon, who ordered some of the original debris that was brought into Roswell by Mack Brazel that day to be flown to Washington for immediate inspection. That flight passed through Fort Worth, where DuBose checked the container, a canvas pouch, holding debris samples prior to sending it on its way to its final destination. He did not look at the contents of the pouch.[1]

The second flight's destination was Wright Field, after a brief stop at Fort Worth. Piloting the plane was Roswell's deputy base commander, Lieutenant Colonel Payne Jennings. Jennings was disliked by just about everyone who came into contact with him, because of his over-the-top military attitude. Enlistees detested and feared him, while fellow officers simply referred to him as "Peter Pain."[2] Jennings, who had a reputation for "hard landings," would lose his life five years later during the Korean War while trying to land a B-29 with a bomb stuck in its bomb bay. The copilot was the base executive officer, Lieutenant Colonel Robert I. Barrowclough. Rounding out the command crew were Major Herb Wunderlich of the First Air Transport Unit (the "Green Hornets") and Captain William E. Anderson of the Air Base Squadron.[3]

All of the noncommissioned officers on the flight were from the 830th Bomb Squadron and included Master Sergeant Robert R. Porter (who was the crew chief), Technical Sergeant William A. Cross, Technical Sergeant George M. Ades, and Technical Sergeant Sterling P. Bone.[4] Also on this flight was Marcel, who had been ordered by Blanchard to accompany the material that he and CIC Captain Sheridan Cavitt had brought back from the Foster ranch earlier that day.

The standard C-54 cargo plane at the base was not used to transport the wreckage. Instead, a B-29 bomber was readied for the unscheduled flight. Although it remains unclear why this choice was made, the choice reinforces the unusual circumstances of this flight.

It has been possible to piece together details about this particular flight from interviews conducted with firsthand witnesses who were either on the plane or on the ground at Roswell or Fort Worth. By comparing and combining their testimonies, one can reach certain conclusions regarding the nature of the debris this second flight was ferrying, and how it might relate to Ramey's press conference and the photographs taken in his office that day.

Upon his arrival, as later related by Roswell PIO First Lieutenant Walter Haut, Marcel carried a box of genuine debris that he had held in his lap on the flight to Ramey's office. The box included the small I-beam that Marcel's son Jesse Jr. would describe years later as displaying indecipherable symbols along its inner surface. Marcel placed the box on Ramey's desk in the general's office. Ramey then directed Marcel into another room to indicate the crash location to him on a large wall map.[5]

When they returned to the main office, Marcel immediately observed that his box of real wreckage had been removed, and the remains of a weather balloon and a torn and mangled radar target were laid out on the floor. A reporter by the name of James Bond Johnson (now deceased) from the *Fort Worth Star-Telegram* was asked to step into the office, and Marcel was then posed for two pictures with the substituted balloon remains, after which he was instructed by Ramey to not say anything to anyone, and that he would handle the entire situation. More pictures were taken of Ramey alone and with DuBose, and the remainder of the flight to Dayton, Ohio, was *officially* cancelled.[6]

What was actually going on, however, was that Marcel had been abruptly removed from the flight to Wright Field and ordered to return to Roswell, while the real wreckage from the B-29 was transferred to another plane to complete the original mission. The resumption of this flight, contrary to Ramey's previous statement, was also confirmed by the local FBI office in Dallas in the now widely circulated telegram dated 6:17 p.m. CST on July 8. Marcel returned to Roswell the next day complaining to Haut about the "staged event" in Fort Worth in which he felt that he had, unfortunately and unwittingly, played a part.

The late Robert R. Porter, the flight engineer on the original flight, confirmed the extraordinary security measures that surrounded every aspect of the assignment. "Whatever was in the cargo hold was escorted by an armed guard who had been assigned to it from Roswell," he said. This would suggest that something extremely important or highly classified was on board.

Porter recalled that three or four shoebox-sized packages wrapped in brown paper and one triangular-shaped package, also wrapped in brown paper and about 2½ to 3 feet across at its base by 4 inches thick, were loaded onto the plane. These had been handed up to him through an open hatch on the B-29 while it was still going through preflight near the operations building. A staff car from Building 1034 had driven up to the plane and delivered the packages, which Porter personally received. All of them were extremely light, and they were stored in the forward section of the plane. Although Porter definitely remembered Marcel on this flight, he did not recall any other debris. Unknown to him, actual material would be loaded just before the aircraft would begin its taxi to departure.[7]

First Lieutenant Robert J. Shirkey was the former assistant operations officer for the 509th and the officer on duty when the July 8 flight to Wright Field taxied up to the flight operations building. He was responsible for drawing up its flight plan.

According to Shirkey, shortly after he had returned from lunch (about 1:15 p.m.), he was informed that a flight plan had to be drawn up for an unscheduled 2 p.m. flight to Wright Field. No sooner was he told this than the plane—a four-engine B-29 bomber—taxied up to Flight Ops for checkout. Shirkey could see some of the crew inside the cockpit. He recognized Lieutenant Colonel Payne Jennings in the pilot's seat but doesn't recall the others.[8]

Just then, he heard a loud voice behind him wanting to know if the flight was ready. Shirkey recognized Blanchard's voice. He replied that it was, and Blanchard stepped out into the hallway and waved to some people who were waiting outside on the street side of the building to come on through.

Blanchard backed up into the doorway to allow the men to pass, and in the process, blocked Shirkey's view of the procession down the hallway.

After asking Blanchard if he could step aside a bit so that he could see some of the action, he found himself standing "buckle to buckle" with the base commander as the men filed by them. There were at least half a dozen wearing dark blue suits ("FBI types"), none of whom he recognized, except for Marcel, who was carrying an open cardboard box filled with non-reflective, aluminum-looking "scrap metal." All but one of the other men were carrying similar open boxes of the same material.

One particular item in Marcel's box caught Shirkey's eye and still stands out in his mind today: "Sticking up in one corner of the box . . . was a small I-beam with hieroglyphic markings on the inner flange, in some kind of weird color, not black, not purple, but a close approximation of the two."

Following Marcel was one of the FBI types, carrying only a single piece of metal under his arm. It was about the size of a "poster drawing board" (about 2 feet by 3 feet). The men moved quickly through the building out to the waiting B-29, and he managed only a brief look. "Here it came, and there it went," he would later remark. The FBI types handed the boxes of material up through an open hatch on the plane, and all clambered aboard the flight.

Former RAAF Lieutenant Robert "Bob" Shirkey in a 2002 photo snapped just as Shirkey had snapped, "Well, are you gonna take it or not?!" (Photo courtesy of Tom Carey.)

Sometime shortly before or after the Marcel troupe made its hallway dash (Shirkey cannot remember which), he recalls an Army staff car driving up to the waiting aircraft, whereupon someone got out and handed a few plain packages up through an open hatchway to someone inside the plane. After the hatch closed, the engines revved up as the B-29 rolled down to the runway and made a speedy takeoff. Blanchard then turned and tossed a perfunctory "see you" in Shirkey's direction as he left the building. He never saw Blanchard again. Nine days later, Shirkey was transferred to the Philippines.[9] He is now deceased.

Even though Porter had been advised by one of the officers on the flight (later identified as Anderson) that the material in the cargo hold of the plane was from a "flying saucer" and that he (Porter) was not to say any more about it, Porter still wasn't sure of its true nature, "... whether it was Brazel's material or something else."

According to Porter, when they landed in Fort Worth, the officers were permitted to disembark, but enlisted personnel were told to remain on board until the plane was secured, meaning that guards were posted around it. Afterward, they were allowed to go to the mess hall to eat, during which time the material was transferred to another plane, a B-25 that would fly it on to Wright Field. When they returned to the B-29 for the return trip to Roswell, they were informed that the material they had flown to Fort Worth under so much secrecy and security was simply a weather balloon.[10]

What about those suspicious packages that were loaded into the plane back in Roswell just before takeoff? Where might they fit into this story? As you will recall, three or four of the packages were of the shoebox variety, and the fourth was triangular in shape with the base edge being 2½ to 3 feet across. All of the packages were wrapped in plain, brown paper and taped shut. Porter doesn't know what became of them after landing in Fort Worth. All he knows is that when the crew returned to the plane after eating a snack before the return trip to Roswell, the packages were missing. He assumed they had been transferred to another plane to be flown on to Wright Field. He was wrong, as it turned out, because there was no "on to Wright Field" intended.[11]

The answer leaps out when you look again at the Fort Worth photographs taken in Ramey's office. Scattered on a blanket of brown wrapping paper in the middle of the general's office, with all of its triangularity on display, is the torn-up tinfoil radar target. Folded, with a broken strut or two, it could easily fit the dimensions of the package received by Porter. Also on display in the pictures of Ramey and DuBose, and especially Newton, is one of the brown-paper-wrapped containers. It can be seen on the floor next to one of the chairs in the Newton photo, and behind the middle chair in the Ramey/DuBose photos.[12]

The July 8 flight from Roswell to Fort Worth was a special flight. Much secrecy and security, if not urgency, surrounded it. The high rank of the crew-members indicates the kind of priority attention that would not be accorded a flight transporting mere rubber and tinfoil. The talk on board was that they were carrying pieces of an actual flying saucer, but the crew was warned to keep their mouths shut about it. If anyone on board knew the real mission of the flight, it would have been the highest-ranking officers, Jennings and Barrowclough, who were close to Blanchard. Marcel didn't know, and neither did Anderson, who also believed the cargo was not made on Earth. It was not until the start of the return flight that the crew was officially told that they had flown a weather balloon, and to forget about it. Similar before and after reactions are described in the July 9 flight to Fort Worth, which we contend transported a number of the bodies. The fix was in; everyone should just go home and act as though nothing happened.

So, if the balloon debris arrived from Roswell, did it originate there? Some have suggested that the White Sands Missile Range near Alamogordo was the source, and others say it was Wright Field or even Fort Worth itself. More likely, however, for the most obvious reason, which flies in the face of those who would maintain that Marcel and Blanchard himself would not have recognized a rawin target balloon, is that the balloon and target did indeed come from the base at Roswell. Why? Because the RAAF was launching such balloons from atop Roswell's tallest building on the average of twice a day in connection with the base's frequent test drops of unarmed atomic bombs.

But one other thing still remains certain. There are absolutely no witnesses to the recovery of a weather balloon or any other balloon device (or its reinvention by the Air Force in 1994, Project Mogul included) at the Foster ranch in June or July of 1947. The weather balloon, which was packaged up and shipped out on July 8, and which would serve as a prop for the official explanation of what actually crashed, came from the RAAF. It was not what Mack Brazel discovered and then brought into town. This scenario, as has been described by the actual witnesses, represents the best explanation available to us that supports the facts as we have come to know them.

13

THE SECRETARY AND THE SPACEMEN

From all eyewitness accounts, something unusual was happening inside the RAAF hospital that July. Unfamiliar doctors and nurses rushed through the halls and into and out of rooms. Regular staff were sent back to their living quarters. No one was talking except for guarded whispers. MPs were posted around the outside perimeter as well as inside along the main emergency corridor. Ambulance trucks would pull up to the rear loading dock area that led directly to the OR. As First Lieutenant Rosemary A. McManus, a nurse with the base medical unit, told us just weeks before passing away, "Something big had happened." She would acknowledge nothing more.[1]

The highly experienced hospital administrator, Lieutenant Colonel Harold M. Warne, had been exposed to the worst of human atrocities as a medic through WWII, and then at the first atomic base in the world. Even in 1947, planes would crash during training exercises, and bodies mangled and burned beyond repair had become all too common. But "something big" had happened here, and apparently it was not described in any medical journal. And even though Warne was in charge of the medical squadron, he was not directing this assignment. Therein may be the cause of his behavior as opportunities would later present themselves.[2]

All military hospital administrators had their own secretary. Warne was no different. His was a twenty-seven-year-old civilian woman by the name of Miriam "Andrea" Bush. Miriam was a graduate of New Mexico State College at Las Cruces, majoring in business administration. She would have graduated at the beginning of WWII, and because college campuses were principal recruiting grounds for the FBI at that time, all indications are that she also had a background in intelligence, which would explain why she would land a top security job at the RAAF after the end of the war in the South Pacific.[3]

Now, one item of crucial importance needs to be emphasized here: the RAAF hospital in 1947 did not have a morgue. That is why the base had a contract with a private mortuary, namely the Ballard Funeral Home. The city of Roswell did not have its own coroner at that time, so they received assistance from the state. All reports of extra security and the presence of outside personnel took place at the precise timing of a reported crash of a genuine flying saucer on that remote ranch north of town. If civilian fatalities were involved, they would have gone directly to one of Roswell's two funeral parlors. If they were military, they would have gone first to the base hospital and then to the private mortuary. Curious phone calls were made to the Ballard Funeral Home inquiring as to the availability of "children's caskets." A rather strange request on the face of it, but even more so coming from a facility without a morgue. Dry ice was called in from Clardy's Dairy during this same period of time. Also, there were follow-up calls to the mortician about recommended embalming techniques that would be the least detrimental to tissue and bodily fluids.[4] You get the picture. It sounds as though the RAAF hospital had some corpses beyond the realm of regular practice standards as regulated by the state coroner's office. And in any event, absence of a morgue notwithstanding, the base hospital would have to do for the time being.[5]

It was dinnertime during one of the days highlighted by all of these strange circumstances when Miriam Bush would arrive at her parents' home from a rather memorable day at the base hospital. She would sit down with her mother and father, who was the first chiropractor to set up a practice in Roswell, and her brother George and sister Jean. Many years later both George and Jean recounted how upset Miriam became, and how she wouldn't touch her food. She then excused herself and started to sob uncontrollably as she raced into her bedroom. Both of them had great respect for her employment at the base. Did she lose her job? Did she lose a close friend? George sensed something worse. "Fear seemed to overcome her," he said. Dr. Bush responded immediately with similar concern for her well-being.[6]

The story she would confide was told between tears and near to shock. It sounded like a nightmare, but her emotional response was too real. It was something she was not prepared for. None of them were. She was eventually able to describe how she had been performing all of her regular duties at the hospital earlier that day, but grew more and more curious as to all the additional personnel who acted totally indifferent to the normal staff. Whether it was out of frustration for being left out of all the commotion or just a desire to share all the excitement with someone, Warne would take her by the arm and quietly mention that she should accompany him to the examination room. Upon entering surroundings that normally would have been quite familiar, she immediately was surprised to observe a number of bodies on gurneys in the middle of the room. But something was wrong. Something became terribly wrong. At first she cried out, "My God! They're children!" But she soon realized that their body size was their only childlike quality. Their skin was grayish to brown in tone and white linens covered most of each body. But the heads, the heads were too large. And the eyes, those large eyes that wouldn't shut. "Those staring eyes," she said. Panic started to quicken her heart, and then it happened: "One of them moved!" All her father could do was listen with total disbelief and hold her as she wept. He was aware of all the talk of a crashed spaceship outside of town and the spacemen inside it. But now it had hit home. And there was nothing he could do about it. Eventually, she would cry herself to sleep, though one might debate whether sleep would serve as any respite.[7]

Morning came all too soon, but Miriam wrestled with her professional training, and her fear grew more and more into anger at her boss, "Why did he have to show me something so terrible?" she thought. "Why did he have to involve me?" But the entire town of Roswell was abuzz with all the talk of the crash of a flying saucer on some ranch and "little men" that were found inside it . . . and some of them were possibly loose in town! Only the base south of town could provide the answers, and only the hospital knew the truth. The morning newspaper carried headlines of it all just being about some old weather balloon. How silly, she thought.[8]

Front view of the Administration Building of the RAAF Base Hospital complex in a late 1940s photo. (Photo courtesy of the U.S. Air Force.)

Much had taken place overnight while Miriam slept. A temporary morgue was set up, a large wooden crate was constructed and packed with dry ice, a tent was pitched, and a metal fence was erected around it on the far south end of the tarmac. In the meantime, most of the activity back at the hospital had returned to normal—as though nothing "big" had ever happened. The day was Wednesday, July 9, and the big shoe was about to drop on Miriam.

As did so many others merely performing their military duties at the RAAF, Miriam became immediately suspect. Any base personnel who saw anything out of the ordinary would have to be warned of the consequences of speaking out of turn, and the traumatized secretary was no exception. Her brother George somberly described to us her demeanor that evening, as she said, "I am never to say another word about what I saw. None of you ever heard me say anything about it," she chided them. According to George and Jean, she displayed all the symptoms of being subjected to heavy-handed threats. She would become more and more paranoid about the entire ordeal. Yet she couldn't share even her worst fears with the very family who also knew the truth. There was nothing any of them could do and certainly nothing any of them could prove. The whole situation became rather hopeless. Best to do just as the military sternly advised—never to say another word.[9]

No one ever questioned her truthfulness, and she never did mention it again. But it had made such a lasting impression on her brother that years later when he would marry Patricia, it was one of the very first private pieces of family history he confided to her. Sadly, no one was ever able to get through to Miriam again. Whatever she saw back in that examination room in 1947 haunted her relentlessly. She would marry within a year—someone she had just met—move to California, and try to forget the unforgettable.

After nearly forty years of a loveless, "arranged" marriage, she would finally file for divorce in 1987. A tremendous weight was lifted off her shoulders; she was not distraught or depressed about the failed relationship. Such was the distinct impression from her sister-in-law Pat, who would speak to her on the phone on a regular basis. Within months of the separation, Pat sensed a subtle change becoming the focus of each new conversation. Miriam was becoming increasingly paranoid. She was deeply concerned about being watched and followed, which to Miriam's sister Jean all seemed to be connected in some way to 1947 and the purpose of Miriam's thirty-nine-year marriage to a homosexual.[10]

The late Patricia Bush would receive one last call from Miriam in December of 1989. She had become obsessed with the fear that someone was spying on her day-to-day activities. Nothing Pat could tell her would alleviate her dread. Still, no one in the family suspected that time was about to run out for Miriam.[11]

The very next day, Miriam would check into a motel just north of San Jose in the town of Fremont, strangely using her sister's name. She was unaccompanied and would not be found until the next morning. The coroner's report concluded that she had committed suicide by wrapping a plastic bag around her head—a rather prolonged and gruesome way to take one's own life. According to the Bush family, scratches and bruises also covered her arms. Other suspicious details suggest that Miriam's fears may not have been totally unfounded. The truth she possessed about Roswell had died with her—death being the great silencer.[12]

When Roswell investigator, Victor Golubic, found Dr. Jack Comstock, the base's chief surgeon back in 1947, just a few years later and asked him to comment on the unearthly visitors at the old RAAF hospital, he denied having any such knowledge. And Miriam Bush? "I have no memory of such a person," he said, denial being the second greatest silencer.[13]

14

"GET THESE OVER TO THE BASE HOSPITAL—NOW!"

The first book ever published about the Roswell Incident dealt mainly with events surrounding the Corona sheep rancher, Mack Brazel, who initiated the Roswell chain of events when he discovered one of his sheep pastures almost totally blanketed with strange wreckage following a severe lightning storm.[1] There was no suggestion that Brazel had found anything other than wreckage—there were no bodies, and the wreckage found on his ranch consisted mostly of very small, very thin pieces of debris suggesting a midair explosion and the notion that perhaps the main part of the stricken craft, whatever it was, had crashed elsewhere, at a second site.

This second crash site was revealed in a story told to close friends and family members by a deceased soil conservation engineer who lived in Socorro, New Mexico, by the name of Grady L. "Barney" Barnett. Before being chased away by the military, Barnett claimed to have seen a downed flying saucer along with its dead crew up close while working on the Plains of San Agustin just to the southwest of the town of Magdalena, 150 miles west of Brazel's sheep pasture. The suggestion was that the UFO, after possibly being struck by lightning near Corona, exploded in the air, raining debris down on the Foster ranch. What was left of the craft then careened out of control to the west, finally falling to earth on the Plains of San Agustin. Although the re-tellers of Barnett's story could not provide a year when Barney's chance encounter was supposed to have occurred, Berlitz and Moore assumed that it must have been 1947 and simply stapled Barnett's account to the Brazel story to close the circle for the Roswell Incident. Barnett was known by everyone to have been a straight shooter, but there were major problems with his account, not least of which were a complete lack of corroborating witnesses, the lack of a date for his encounter, and the distance of the alleged "Plains of San Agustin" crash site from Roswell, which was much too far for the RAAF to have become involved in any recovery operation there.

By 1991, new first- and secondhand witnesses to alien bodies had been added to the expanding Roswell story, but the location of the supposed second site came into question. In their book, the investigative team of Kevin Randle and Don Schmitt, who had reopened the investigation of the Roswell case three years earlier, kept the Barney Barnett story in their Roswell crash scenario, but moved Barnett's close encounter from the Plains of San Agustin 152 miles east, to a low bluff about two miles east/southeast of Brazel's pasture.[2] The move was based upon information the team had gathered from local Corona ranchers of heavy military activity at that location and time frame. They then speculated that Barnett must have been in Lincoln County and not Catron County on the day of his discovery. There was no evidence of that, but it was the only way to include Barnett's alleged discovery that there were indeed bodies recovered from the UFO crash, but that they were recovered near the Brazel debris field. With three new witness accounts of alien bodies—Sergeant Melvin E. Brown, the RAAF cook from K Company who accompanied the bodies from the crash site to the Roswell base in the back of a truck;[3] Captain Oliver W. "Pappy" Henderson, the First ATU pilot who flew wreckage and bodies from Roswell to Wright Field in Dayton, Ohio;[4] and the Roswell mortician, Glenn Dennis, who claimed to have inadvertently arrived at the Roswell base hospital during an attempted "alien autopsy"[5]—all signs pointed away from the Plains of San Agustin. In 1994, Randle and Schmitt dropped the Barney Barnett story altogether from their Roswell UFO crash scenario and again moved the final crash site—now termed the "impact site"—to a site much closer to Roswell.[6] This site—known as the Corn Ranch Site or the Kaufmann Site—later turned out to be bogus, as its location was based upon the testimony of a single, alleged eyewitness who himself was later discovered to have been a purveyor of false information.[7]

On the morning Mack Brazel was making history in his Corona sheep pasture, he was accompanied by a seven-year-old neighbor boy, Dee Proctor, who often visited Brazel from an adjoining ranch owned by his parents, Floyd and Loretta Proctor. In the ensuing fifty-nine years up to his death in 2006, Dee Proctor never allowed himself to be formally interviewed by anyone about what he witnessed with Brazel that day so long ago. In 1994, however, he took his dying mother to the second site.[8] This site turned out to be the exact same site identified by Randle and Schmitt as the "impact site" in 1991. Today, we believe that the "Dee Proctor Body Site" is a legitimate site, the second in the straight-line trajectory of three sites involved in the Roswell UFO crash, where Mack Brazel found two or three alien bodies a few days after they were blown out or ejected from the stricken craft when it exploded over the debris field on the evening of July 3, 1947.

The "Dee Proctor Body Site," a low bluff on the J. B. Foster Ranch where two alien occupants of the stricken UFO were ejected and fell to earth and their deaths. (Photo courtesy of Tom Carey.)

The rest of the ship or an escape vehicle, along with the remainder of the doomed crew, remained airborne and continued in an east/southeast direction for another thirty miles before crashing in a flat area with low, rolling hills forty miles north/northwest of Roswell. Based on new evidence, we located the third and final crash site—the true impact site—in 2005. It was at this last site, about which Mack Brazel knew nothing, that an additional two or three dead aliens and one live one were discovered by civilian archaeologists. The following witness accounts presented in this chapter all pertain to recovery activities that took place on the Foster ranch at or near the Dee Proctor Site during the July 7–8, 1947, time frame.

In July of 1947, Ed Sain was a private first class in the 390th Air Service Squadron attached to the 509th Bomb Group at the RAAF. As did most others in the 390th ASS, PFC Sain possessed the top secret security clearance that was required for security personnel whose main duty was to guard the Silverplates. Sain was just about to turn in for the evening (July 7) when the chief of security, Major Robert Darden, burst into the barracks: "C'mon, boys! We've had a crash." Sain and a 390th ASS buddy of his, Corporal Raymond Van Why, were told to report to the ambulance pool outside of the base hospital ASAP. After the short walk to the hospital, the two airmen were directed to a waiting "box-type" military ambulance, which they quickly entered. They drove north of town for half an hour or so, then headed west "into the boondocks" of Lincoln County. Because it was dark outside, they could not see where they were going or where they had been. The ambulance finally came to a stop in the lee of a small bluff, around which there had been a beehive of activity just a few hours before. Except for a few tents that had been erected at the base of the bluff and a number of floodlights that had been set up, there wasn't much to see but desert. "Major Darden and Major Easley [head of the 1395th MP Company] were both there, which was unusual," Sain told us. Sain and Van Why were each given a handheld searchlight and told to guard the entrance to the site from a tent set up for that purpose. Their orders were to "Shoot anyone that tries to get in!" According to Sain, "We had plenty of food to last us the night and thank God that no one showed up. So we didn't have to shoot anyone. We were relieved before first light and back on the base at daybreak."[9] Due to a combination of Ed Sain's age and his accent, it was difficult to understand him over the phone at times. We also felt that he was not giving us the whole story regarding that night in the desert. A call to Sain's son Steven (with whom we had talked prior to speaking with his father) confirmed our suspicions. It had taken Steven Sain and his brother almost thirty years to get anything out of their father regarding Roswell. "He was extremely reluctant to talk about it," Steven said. "He said that he was under a security oath and feared for his life if he said anything. He wouldn't talk about it for a long time, and he still won't watch any TV shows or read any books about Roswell. It's only been recently, however, that he has started talking to us."

According to Steven Sain, his father told him that he had been an MP at the RAAF at the time of the Roswell Incident. His job had been to "guard the bodies at the crash site" by guarding the entrance to the site with another fellow. "He referred to the bodies as 'little green men,'" Steven said, "when he told me, and said that they were kept in one of the other tents until being transported to the base. My father said that everything was strange that night, and he must have seen the ship at some point, because he told me that it was the strangest thing he had ever seen in his life."[10]

Raymond Van Why passed away in 2001 at the age of seventy-six. According to his widow, Leola, her husband was usually pretty closedmouthed about the nine and a half years he spent in the military. "He was a security guard who guarded the Enola Gay, and when he got out of the Air Force, he shredded his service records." According to Mrs. Van Why, her husband first talked about Roswell after he got out of the service in 1954 when, upon reading an account in the newspaper or a magazine of an alleged spaceship crash, he shouted out, "I saw that!" He then told her that, when he was stationed at Roswell a few years earlier, he had been a guard at a crash site "out in the desert" where a spaceship had crashed. "My husband told me that it was a UFO that had crashed, that it was a round disc." We then pressed her on that point: "How did he know that?" She answered, "Because he was out there and saw it!"[11]

Van Why wasn't the only one. Sergeant LeRoy Wallace, a 6-foot 9-inch Cherokee Indian from Arkansas, was assigned to the 390th ASS as an MP three months prior to the Roswell Incident. According to his widow, he was called away one evening to go to a crash site outside of Corona "to help load the bodies." When her husband returned home early the next morning, the first thing she noticed, besides his disheveled appearance, was the smell. "The stench on his clothes was the worst smell you'd ever want to smell. It was worse than any combination of smells you could imagine. I had him strip off his clothes. We ultimately burned them and buried the ashes." She said that her husband bathed frequently with lye soap and an old Army scrub brush after that, and would wash his hands up to ten times a day to the point of becoming raw. "He walked around for two days after he returned home and did not sleep, and for the next two weeks, when he ate, he wore gloves because the smell was still on his body." Wallace was transferred out of Roswell three months later.[12]

The witnesses keep coming. After World War II, twenty-six-year-old Sergeant Frederick Benthal was serving in the Army Air Forces as a photographic specialist at the Anacostia Naval Air Station in Washington, D.C. In 1946, he had helped set up the photographic equipment for Operation Crossroads in the Pacific, involving special cameras for filming the two atomic bomb tests conducted there. According to Benthal, after reporting to work one morning in early July of 1947 with a friend, Corporal Al Kirkpatrick, they were told to pack their bags for a flight to Roswell, New Mexico. They flew in a B-25 *Mitchell* medium bomber, leaving around 10 a.m., and made one stop along the way, during which they were told not to leave the plane. On the flight, the two men studied the dossiers of persons who might be expected to be at their destination. These individuals included J. Robert Oppenheimer of Los Alamos ("Father of the Atomic Bomb") and General Curtis "Bomber" LeMay of WWII fame and the future head of the Strategic Air Command and the Air Force's Chief of Staff. The plane landed in Roswell around 5 p.m., and the men checked in to the base transit barracks for the evening. The following morning, Benthal and Kirkpatrick were picked up by a covered military truck and headed north of town. During the trip, both men changed into rubberized suits that were very hot, but apparently offered some kind of protection—protection against what, they did not know.

When they arrived at the site, Benthal saw several tents that had been set up near a small bluff, and what appeared to be a refrigerator truck. He also witnessed covered trucks leaving the site that were obviously carrying wreckage of some sort. He could see thin strips of wreckage sticking out of the backs of the trucks as they departed. Other empty trucks continued to arrive at the site. Benthal witnessed a lot of enlisted men going back and forth in various directions, as well as two majors whose names he did not know. He and Kirkpatrick were then split up, as Kirkpatrick was ordered into one of the empty trucks that headed out to another location (the Brazel debris field site), while Benthal was taken to a nearby tent and told to stand by. An officer then came out of the tent and told Benthal, "Get your camera ready!" Then the officer looked into the tent and made a loud comment that someone was coming in, whereupon a number of officers then exited the rear of the tent. Once inside, the officer told Benthal to stand back and then pulled back a tarp that was on the floor of the tent, revealing several little bodies lying on a rubber sheet. Benthal and the officer slowly but purposefully moved around in a circle, with Benthal taking pictures of the bodies lying in death beneath them. "They [the bodies] were all just about identical, with dark complexions, thin and with large heads," Benthal said. "There was a strange smell inside the tent that smelled something like formaldehyde."

Benthal was shooting his pictures with a standard-issue Speed Graphic camera that had a holder, each with two shots. Although it was daylight outside, it was dark inside the closed, rubber-lined tent. Because it was so dark inside the tent, the flash bulbs gave off a blinding light when a picture was snapped. After taking each set of two pictures, Benthal would give the holder to the officer. This procedure was repeated many times as the two men circled the bodies. The entire session lasted about two hours, after which Benthal was told to leave. About that time, Kirkpatrick returned from the other site in a truck that Benthal could see was loaded down with wreckage. Benthal and Kirkpatrick were dismissed from the site and returned to the base in Roswell. They were debriefed on the ride back to the base and told not to talk to anyone. "My camera case, cameras, and all of the film had been confiscated before we left the site. We were given bunks in the barracks overnight, in a small room upstairs, rather than in the larger room (downstairs) that had many more bunks."

The men were awakened around 4 a.m. the next morning and taken to the mess hall where they ate the early-bird breakfast. After breakfast, they boarded the B-25 and headed back to Washington. When they got back to Anacostia, they were again debriefed, this time by a Marine officer, a lieutenant colonel by the name of Bibbey, who asked them if they knew what they had photographed. Benthal and Kirkpatrick both responded, "Yes, Sir." To which Lieutenant Colonel Bibbey instructed them that they did *not* know what they had photographed! Then Lieutenant Colonel Bibbey asked the question again. Both men responded with a resounding, "No, Sir!" and then were summarily dismissed.

Recalling the episode in 1993, Benthal observed, "Not long after that, I was assigned to the Arctic to take pictures of pieces of equipment to study the effects of cold [on them]."[13]

One of the MPs Frederick Benthal might have seen guarding the crash site was Corporal William Warnke of the 390th ASS. Warnke, now ninety-three years old, was normally assigned to guard the Silverplates on the flight line at the base, but he recalled being sent "out in the boondocks" in July of 1947 to stand guard.[14] Another might have been Corporal Leo Ellingsworth, who in July 1947 was a member of the 830th Bomb Squadron at the RAAF. In 1950, as Ellingsworth was about to be shipped out of the country, he told his brother Ross a story that Ross did not know whether to believe or not. Leo Ellingsworth passed away years ago, but fortunately he also told the story to his niece, Monte Dalton, before he died. According to her, Ellingsworth had been assigned to a detail one night with a lot of other men who were ordered to a location "out in the boondocks" to pick up "bits and pieces of material." According to him, the men "walked shoulder-to-shoulder all night long," back and forth across the site, picking up whatever they could find and placing it in a container that each man had been given. When they were finished, each man emptied the contents of his container into a wheelbarrow as he departed the site. "My uncle also told me that he saw 'three little men,'" she said.[15]

It was the late Roswell embalmer, Glenn Dennis who first introduced "smelly" alien bodies at the RAAF base hospital into the Roswell story. Dennis began talking publicly about it in the late 1980s and early 1990s, and has repeated his story in numerous documentaries about Roswell since then. Often referred to as the "Roswell mortician" for viewing audiences, Dennis was a twenty-two-year-old embalmer employed at the Ballard Funeral Home in downtown Roswell. On the morning of July 8, 1947, Dennis claimed to have received a series of telephone calls from the mortuary officer at the base, first inquiring about the availability of child-size caskets, then later about various embalming techniques for bodies that had been "out in the desert" for an extended period of time. Thinking that

Roswell mortician Glenn Dennis in a 2003 photo. (Photo courtesy of Tom Carey.)

there might have been a plane crash—not unknown to Roswellians with the air base so close to town—Dennis offered his assistance. It was declined on the premise that the calls to Dennis were simply for "future reference just in case." Later that afternoon, Dennis received a call instructing him to pick up and transport to the base hospital a local airman who had just been injured in a motorcycle accident in town (besides providing mortuary services to the base, the Ballard Funeral Home also had the base contract for ambulance services when requested). Upon arriving at the emergency wing of the base hospital, Dennis got out of his vehicle and proceeded to walk with the injured airman up the ramp and into the rear of the building. Along the way, Dennis passed several box-type military ambulances that were parked close to the hospital. One of them had its back door swung open, and Dennis could see that it contained what appeared to be the front end of a canoe-like structure with strange writing or symbols along its side. It appeared to be made of metal, and its metallic surface had a bluish hue to it, as if it had been subjected to intense heat.

Once inside the hospital, Dennis then claimed to have bumped into a nurse he knew who was coming out of a side room holding a towel over her nose and mouth. Seeing Dennis, she exhorted to him, "What are you doing here? Get out! Get out of here as quickly as you can, or you will be in a lot of trouble!" On his way out of the building, Dennis was allegedly accosted by a black sergeant and a red-haired captain who called Dennis an "SOB," told him that there had been no crash, that he had seen nothing, and that if Dennis didn't get out of there right away and keep his mouth shut, he might require the services of an embalmer himself. It was a shaken and angry Glenn Dennis who drove back to Ballard's, followed all the way by an armed military escort, just to make sure. A day or two later, Dennis met up with his nurse friend at the Officers' Club on the base to try to find out what had happened at the hospital. Dennis could tell right away that she was nervous and not feeling well. They ordered lunch and Cokes, but neither touched their food, as the nurse tried to describe the scene at the hospital. She told Dennis that she had gone into a side room to get some supplies when a military doctor whom she had never seen before turned around from what he was doing and told her to stop what she was doing and come over, that she was needed at once to take notes. It was then that she noticed the overwhelmingly foul smell in the room and three dead "foreign bodies" lying on gurneys. Two of them were in bad shape and appeared mutilated, while the third one seemed to be intact. It was over the intact one that several doctors were hovering. She described the creatures as being 3½ to 4 feet tall, with long arms and frail bodies. They had oversized heads, and the bones of the head were similar to those of a newborn baby, meaning that they were pliable and could be moved slightly. Each hand had only four fingers—no thumbs—with suction-cups on the tips of each finger. The eyes were sunken and oddly spaced. The ears and nose were simply holes in the side of the head and on the face, and the mouth was only a small slit. She did not notice what, if anything, they were wearing, as she was overcome by the horrible smell and became sick. She asked to be excused and left the room, and that was when she had run into Dennis in the hallway. She said that the doctors also became sick from the smell and had to abort the attempted autopsy. The nurse then took out a prescription pad and drew on it a picture of what she could remember that the "foreign bodies" looked like. She gave the drawing to Dennis and excused herself because she was still feeling sick. Dennis never saw the nurse again. According to Dennis, she was transferred to England a week or two later, and a letter that he sent to her a few months afterward came back simply marked DECEASED. Dennis told Roswell investigators that he later learned that the nurse had been killed in a plane crash during a training exercise. He also said her name was Naomi Selff.

After several years of searching without success for any record of a military nurse by the name of Naomi Selff, or any record of a plane crash in England involving American nurses during the pertinent time frame, Glenn Dennis was confronted by Roswell investigators with this information in the mid-1990s.

His surprising and disappointing response was, "That wasn't her real name. I gave you a phony name, because I promised her that I would never reveal it to anyone."[16] In any court of law, when someone is caught in a lie, that person is said to have been "impeached" as a witness, meaning that his or her testimony, as evidence, cannot be relied upon. Even though we know of witnesses who have told us that Dennis had told them about the phone calls from the base for child-size caskets way back when they happened, and of witnesses who have told us that Dennis had told them about his run-in at the base hospital long before Roswell became a household word—still, Dennis was found to have knowingly provided false information to investigators and must technically stand impeached as a Roswell witness. There is no way to get around that fact without believable, clarifying information from Dennis himself. Up to his passing, no such information was forthcoming from Dennis.

We continue, however, to investigate the alleged activities that took place at the RAAF base hospital during the time of the Roswell Incident. We have identified a number of first- and secondhand witnesses who have attested to activities suggesting that something highly unusual indeed was taking place there. One such person was Corporal Arthur Fluery. In July of 1947, Fluery and a buddy of his, Paul Camerato, had just transferred over from the motor pool to the base medical unit (Squadron M) a month or two earlier to become ambulance drivers. At the time of the incident, Fluery was assigned to provide special transportation between the airstrip and the base hospital for an expected increase in the number of people flying in and out of the base. This Fluery did. "There were doctors, both military and civilian," he said, "flying in from all over the world it seemed, at least from all over the country, whom I picked up and drove over to the base hospital. I never saw so many planes coming in and going out in such a short span of time. I didn't know what was going on, but I had heard the rumors. My job was simply to drive the ones coming into the hospital, and drive the ones going out to the airstrip."[17]

Barbara Perez, a radio talk show host in El Paso, Texas, during an on-air interview in 2007 with Julie Shuster (the former director of the International UFO Museum & Research Center in Roswell), stated that her mother had worked at the RAAF base hospital at the time of the 1947 UFO incident and had told her about the extra security at the hospital when it happened. No doubt. Once the bodies were discovered, the base hospital became the center of attention and security until they were flown out to another facility.

One of the chief complaints by skeptics of "The Glenn Dennis Story" has to do with Dennis's claim of having received phone calls from the RAAF "mortuary officer" wanting to know about the availability of "child-size" caskets. Skeptics claim that Dennis only started talking about it in the late 1980s and early 1990s when the Roswell crash became known to the public—in other words, he was making it up. However, recent support for Glenn Dennis's account regarding the caskets comes from several sources. Adam Dutchover was just a small boy at the time of the Roswell Incident who had a "job" at Dryke Bealor's small grocery store, which included an "in-house" coffee nook, located in the tiny town of Hagerman, just a stone's throw from Roswell. Little Adam Dutchover would occasionally serve coffee to customers who had stopped into the store to buy something to eat and to chat for awhile. Included among the regulars were members of the Roswell Police Department and one or two employees of the Ballard Funeral Home. Dutchover remembers that at the time of the incident all of the talk in the nook was about the *flying saucer* crash and the "little bodies." Although he cannot now recall all of their names, he especially recalls the Ballard's people talking about the need for ordering "small caskets" for the Air Force.[18] Garner Mason of Hagerman, whose father, grandfather, and uncles ran a family mortuary services business at the time, says it was their family business that actually made the delivery of the caskets to the RAAF base hospital. "We received the call from Ballard's, because they didn't have enough of them to fill the order. So we made the delivery to the base. They were actually made out of cardboard."[19]

Rex Alcorn, a long-time friend of Glenn Dennis, remembers Dennis "telling me at the time of the incident about receiving calls from the base inquiring about 'child-size caskets.'"[20]

So did Clifford Butts and William Burkstaller, other friends of Dennis, "back when it happened."[21]

Richard L. Bean was a ninety-one-year-old Roswell attorney with his own law firm and still going into the office several times a week when we met with him in 2004 to conduct a wide-ranging discussion about what he remembered about the 1947 events as they were happening. He told us that he heard about the crashed UFO "within days of the crash," but that he didn't hear Glenn Dennis or Walter Haut talk about it until a year or two later. It was at that time that he heard Dennis talk about receiving calls for children's caskets from the base.[22]

A former MP at the RAAF, L. M. Hall was a motorcycle police officer whose beat in 1947 was South Main Street, Roswell, between the town and the RAAF. Hall had gotten to know Glenn Dennis when he, similar to other police officers, would sometimes take their breaks in the small lounge at the Ballard Funeral Home where Dennis worked. He had also known Dennis when he was an MP at the base, and Dennis would make ambulance calls to the base under the contract that Ballard's had with the RAAF to provide ambulance services. As Hall recalled it years later:

> One day in July 1947, I was at Ballard's on a break, and Glenn and I were in the driveway "batting the breeze." I was sitting on my motorcycle, and Glenn stood nearby. He remarked, "I had a funny call from the base. They wanted to know if we had several baby caskets." Then he started laughing and said, "I asked what for, and they said they wanted to bury [or ship] those aliens," something to that effect. I thought it was one of those "gotcha!" jokes, so I didn't bite. He never said anything else about it, and I didn't either.[23]

Beverly Otto lived and worked for the National Institutes of Health (NIH) in Washington, D.C. She left Washington for Albuquerque in 1985, where she was residing in retirement when we caught up with her in 2008 through a third party. She told us that when she worked at the NIH in Washington, she was part of a group of five or six women who would occasionally get together for dinner at a local restaurant. After all the years, she remembers one occasion especially. According to her, "It was soon after the 'Roswell business' when we were joined for dinner by a 'friend of a friend' in our group. She said that she was an Army nurse and had recently been assigned to Walter Reed Hospital. As usual, we were talking about everything over dinner when, at some point in the conversation, the nurse told us that she had come from the base near Roswell. That caught our attention, and one of us jokingly asked her if she saw 'the little people' there. Then she said, 'Oh, yes! I was the nurse who ordered the children's coffins, because they were just big enough for the *little guys who were in the spaceship*.' She didn't describe them or say how many

there were, but she told us that, right after that, she was called in and transferred." We then asked Mrs. Otto the big question, whether she remembered
the nurse's name. "Heavens, no! All I can tell you is that she appeared to be in
her thirties. We never saw her again after that night."[24]

The other main Glenn Dennis "issue" for the skeptics had to do with "the
nurse" who supposedly had accosted him in a hallway of the RAAF base
hospital during an apparent attempted "alien autopsy" there. We have
already stated that Dennis has impeached himself on the issue by knowingly supplying investigators with a false name for his nurse friend. To us,
his reasoning for this—"I promised her that I would never reveal her
name"—is not persuasive, especially if she is now long dead. Throughout the
years, Dennis did provide us with hints as to who his nurse might have
been: ". . . she was short, had olive skin, and looked like a young Audrey
Hepburn." This description would seem to fit a young RAAF nurse by the
name of Eileen Fanton, whose picture is indeed in the 1947 Roswell base
yearbook.

According to her Air Force record, she arrived at the RAAF in early 1947
but left the base in September 1947 due to health reasons. Further, although
Fanton did not perish in a plane crash, as Dennis had often stated, she did die
at a very young age in 1955, but due to idiosyncratic health reasons.[25] Nev-

RAAF nurse Lieutenant Eileen
Fanton. (Photo courtesy of the U.S.
Air Force.)

ertheless, we located and interviewed two
gentlemen who knew Glenn Dennis back
then. Bob Wolf, a minority owner of radio
station KGFL, told us that he knew Glenn
Dennis and that it was known to him that
Dennis was seeing a nurse on the base
that fit the general description of Eileen
Fanton.[26] LeRoy Lang, a sergeant at the
RAAF in 1947, remembered training Lieutenant Fanton in the proper use of firearms.
He also remembered seeing Fanton and
Dennis together on the base on a number
of occasions.[27] So, after all is said and done,
was Lieutenant Eileen Fanton "the nurse"?
We don't think so.

In 1998, a call came in to the J. Allen Hynek Center for UFO Studies in Chicago from a couple living in Hooks, Texas, who claimed to have information about the "Roswell nurse." Tom Carey was at the time on the board of directors at CUFOS and the person responsible for the organization's Roswell investigation. The call went to him for a follow-up. Mr. and Mrs. Charles "Chuck" Huttanus were interviewed over the telephone individually, and both told the same story: In 1960 Mrs. Huttanus worked in a civilian capacity at Walker Air Force Base in Roswell (the successor to the former Roswell Army Air Field). According to the account, one of Mrs. Huttanus's coworkers, a woman perhaps in her forties, told her that she "was here when they brought the aliens in." The startled Mrs. Huttanus asked the woman what she was talking about. "Back in 1947, I was a nurse and happened to be at the hospital when the aliens from the UFO crash were brought in."

The woman apparently did not elaborate, as the Huttanuses didn't have answers to follow-up questions from Carey. They also refused to provide the woman's name. "She told me not to tell anyone, that she could get into trouble. So, I promised. She also told me that, if anyone ever asked her, she would deny the story." We now had a very interesting story, but, without a contact name, no place to go with it. Or maybe we did. During the discussion with Mr. Huttanus, before his wife joined in, he let it slip that the woman's husband was a long-standing "golf pro at NMMI" (New Mexico Military Institute, in Roswell). In short order, we were able to identify the elderly golf pro as a gentleman by the name of Jim Lowe, and his wife was named Mary, a former nurse. Feeling that this was not a lead we wanted to contact by telephone,

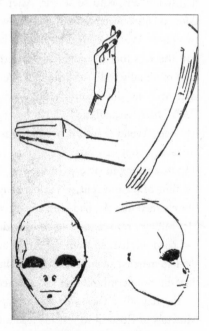

Alleged drawing of an alien body by an RAAF nurse as described to Roswell embalmer Glenn Dennis by Walter Henn. (Drawing courtesy of Don Schmitt/IUFOM&RC.)

we decided to wait until our next trip to Roswell, when we could make an in-person approach. In the meantime, we contacted UFO historian Wendy Connors who lived in Albuquerque to see if she could come to Roswell to interview Mary Lowe, because we felt that a female might have better success in gaining the trust of the witness and getting her to talk. Wendy agreed, and we all met down in Roswell in July 1999. After briefing Wendy on the situation and who the witness might be, we drove her to a location around the corner from the witness's home and dropped her off. The "cold-call" and "ambush" interviews are the most difficult type of interview to undertake, because one never knows how the witness will react at being surprised with something he or she might not want to talk about. Will the witness become hostile and slam the door in your face, or do something else? One never knows, but in certain cases, it's a chance that must be taken. Because, by our calculations, the Lowes had to be in their eighties, we felt that an overtly hostile reaction on their part would be highly unlikely. And so it was. After Wendy Connors knocked on the front door and introduced herself, Mary Lowe invited her into her home. After about two hours, Wendy emerged from the house. Given the length of the interview and the fact that Wendy wasn't bleeding, we felt that it must have gone at least minimally well. According to Wendy, after introductions and pleasantries were over and the discussion got down to Mrs. Lowe's alleged presence at the RAAF base hospital in 1947, her first reaction was to question if Glenn Dennis had been Wendy's informant. She was also very interested in Dennis's sworn affidavit, a copy of which Wendy had brought along to the interview. She read it in its entirety in silence. In the end, Mary Lowe denied being in Roswell at the time of the incident, just as the couple in Hooks, Texas, had said she would. She claimed that she had been an Army nurse stationed in Scotland at the time. Her military records said something different, however. They showed that she had been discharged from the Army in 1946—in the United States—because she had married an enlisted man. At that time in our military, protocols of conduct did not permit officers and enlisted personnel to marry each other.

The day after the interview, we approached Glenn Dennis at the UFO Museum in Roswell to test his reaction to our having found and interviewed the "new candidate" for his long-lost nurse. When he asked who it was, all we gave him was her first name, to which he responded with wide-eyed alacrity, "Oh, Mary Lowe. Yeah, she knows everything." Dennis then seemed to realize that we were watching him closely, as he "caught himself" and looked away. That was all he said, and he left the room. Now, keep in mind, this was a name that Dennis had *never* mentioned to anyone before, and he had nothing more to say about it? Something wasn't adding up. The very next day, we stopped in at the UFO Museum to check on something when a pale-looking Glenn Dennis accosted us and quickly directed us into his office. After closing the door behind us, it was as if he couldn't wait to get the words out: "About yesterday, forget what I said about Mary Lowe. I was mistaken. She doesn't know anything!"

It was obvious to us, then and now, that Mary Lowe believed that Glenn Dennis had been the source of the information about her that Wendy Connors confronted her with during the interview. More than likely, Lowe's first post-interview move was to pick up the phone and call Dennis and read him the riot act. That would explain Dennis's comments to us the following day, delivered with that taken-to-the-woodshed look on his face. It was also obvious that Mary Lowe had not told the truth concerning her whereabouts at the time of the incident. Or did she just forget? And would the couple in Hooks, Texas, just simply make up a story like that out of the blue about a former friend and coworker? We think not. In the ensuing years, Mary Lowe dodged every attempted follow-up meeting with us. Now, because Glenn Dennis is no longer accepting requests for interviews, and the Lowes have disappeared and are presumed to have passed away, we may never achieve a final resolution regarding nurse Mary Lowe's alleged participation in the Roswell Incident. However, with additional leads in hand and new avenues to follow, we are still pursuing "the nurse."

Another man who himself *admits* to having seen the bodies was Eleazar Benavidez (a pseudonym). His wife had told us as much, but it would be several years before he would tell us himself. As a retired career Air Force man, Benavidez feared for his pension—if he ever talked about Roswell. Benavidez and his wife had come into the International UFO Museum & Research Center in Roswell in 2002 to see the exhibits, as did the one million or so other visitors who had preceded them since the museum opened its doors in 1991. Against her husband's wishes, Mrs. Benavidez stopped into the museum director's office to tell someone about her husband. After hearing Mrs. Benavidez's story, especially the part about bodies and the hospital, Julie Shuster, the museum director, felt that a private interview with Mr. Benavidez was warranted. It was soon apparent, however, that there was a major problem: Mr. Benavidez was nowhere to be found. He had left the building when he saw his wife and Julie Shuster looking around for him—a reluctant witness, to put it mildly. Since that day, we managed to track down the former member of the 390th ASS and meet with him half a dozen times to try to gain his confidence. In the process, we learned that Benavidez's main fear, similar to so many others who were and are reluctant Roswell witnesses, was his belief that he, as a retired Air Force veteran, might lose his pension if he said anything about those long-ago events.

We assured Benavidez that we knew of no instances whatsoever of someone losing his pension for talking about Roswell. Besides, the secretary of the Air Force issued a proclamation in 1994 that absolved anyone with knowledge about the Roswell Incident who believed that they were still subject to security or secrecy oaths regarding the matter. The other thing we noticed was that Benavidez was still deeply moved, if not troubled, by what he witnessed back in 1947, and he broke down in tears on several occasions when talking to us about it. His wife also revealed that her husband still has trouble sleeping comfortably and would for years wake up suddenly in the middle of the night, shaking. In 2005, the eighty-year-old Eleazar Benavidez finally agreed to tell his story. He subsequently appeared as a featured witness on the *Sci Fi Investigates—Roswell* TV show that first aired on November 8, 2006.

Eleazar Benavidez in a 2015 photo.
(Photo courtesy of Tom Carey.)

In July of 1947, Eleazar Benavidez was a private first class with the 390th ASS at the RAAF. Possessing a top secret clearance, PFC Benavidez was authorized to provide security support for the most highly classified operations of the 509th Bomb Group. In addition to his primary job of guarding the Silverplates, one of his secondary duties was that of a recovery specialist, which involved the grim activities associated with the aftermath of plane crashes.

The morning of Monday, July 7, 1947, found PFC Benavidez walking back to his barracks after a night of guard duty on the flight line followed by breakfast at the chow hall. Just minutes away, sleep was awaiting him. "Something's going on …" he thought to himself, as he stood at attention and saluted the playing of the national anthem and the morning flag-raising ritual at base headquarters on the south end of the esplanade. He knew that the base commander, Colonel William Blanchard, normally held his weekly staff meetings on Tuesday mornings, but on this day Benavidez thought there were way too many staff cars and other vehicles parked in the headquarters parking lot for a regu-

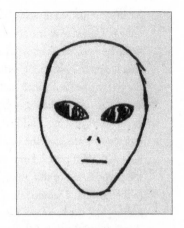

Forensic drawing of the alien witnessed by Eleazar Benavidez by coauthor Don Schmitt 2005. (Photo courtesy of Don Schmitt.)

lar staff meeting. When Benavidez finally arrived back at his barracks, "Word was given to my squadron to be on the alert for special duty," he said. Such was life in the 509th and the Strategic Air Command, and sleep would have to remain a secondary consideration. The word finally found PFC Benavidez: "Benavidez! Get your gun and report to Hangar P-3 for guard duty." Benavidez gave us the following account:

> I got myself ready, got my gun, and reported to the big hangar, as ordered. As near as I can recall, it was late afternoon or early evening at the time. While looking for my OIC [officer in charge] to get instructions for duties at the hangar, I came upon a commotion taking place at the main entrance to the hangar. Some MPs were trying to subdue an out-of-control officer who, among other things, appeared to be drunk as a skunk. I found out later that the officer in question was from my squadron and was the very officer to whom I was to report for a special detail. This officer—whose name I cannot now recall—was to have overseen the transfer of several "top secret items" from the big hangar to the base hospital, and I was there to help escort the transfer. I was later told that he had been to the crash site and had seen the ship. When this officer reported to the hangar and saw the small bodies, it was apparently too much for him to handle, and he just lost it. At this point,

having just arrived myself, a major or lieutenant colonel came out of the hangar, looked over the situation, and pointed at me. "You! Come over here," he said. "You're now in charge of this detail. Get these over to the base hospital—NOW!" He then pointed to three or four gurneys inside the hangar, each of which had something on it that was covered with a sheet. On one of the gurneys, whatever was under the sheet appeared to me to be moving. I saluted my acceptance and understanding of his order, and instructed the rest of the men in the detail to load the gurneys with their payload into the back of a truck that had just arrived for the purpose. Up to this point, I had no idea what we were transporting to the hospital. I would know soon enough, however. As the men were loading the truck, one of the gurneys slipped during the handoff, and the sheet covering it fell away, revealing the grayish face and swollen, hairless head of a species that I realized was not human. My orders were to deliver these to the base hospital's emergency room [Building 317] and remain there until relieved.

When we interviewed him, almost sixty-two years after the event, the fog of time prevented Eleazar Benavidez from recalling the names or faces of the other men assigned to "escort duty" that day. They may well have been from other squadrons on the base or even from other bases brought in from the outside (known as "augmentation troops") to prevent the comparing of notes later on. Benavidez continued:

Upon arriving at the emergency room ramp, we proceeded to unload. I went in with the first gurney and stood aside near the doorway as the medical people took control of the gurney. A half-dozen or so medical and nonmedical officers quickly removed the covering sheet. I couldn't see too well from where I was standing because of the number of officers gathered around the gurney, but I could see well enough to make out that a very small person with an egg-shaped head that was oversized for its body was lying on the gurney. The only facial features that stick out in my mind now are that it had slanted eyes, two holes where its nose should have been, and a small slit where its mouth should have been. I think it was alive. The medical people were mostly just staring at it, but I'm not sure. After the rest of the gurneys were brought into the room, I was dismissed and told to return to my squadron, which I did. There, I was debriefed and made to sign a nondisclosure statement regarding what had just taken place. I was told that if I ever spoke about it, something bad would happen, not only to me, but also to my family. I heard later that the one species that was still alive was apparently taken to Alamogordo, then shipped to Texas or Ohio.

For her part, Mrs. Benavidez confirmed her husband's account of his involvement in the Roswell events of July 1947. She further stated that they have been married since 1949, when he first confided the story to her, and that his memories of that night—seeing the swollen faces and slanted eyes of the "species"—remain with and haunt her husband to this day.[28]

After telling his story before the *Sci Fi Investigates* cameras in the big hangar on the old Roswell base where it all happened back in 1947, all the emotional Eleazar Benavidez could say after the cameras stopped filming was, "Do you think I will lose my pension?"

POSTSCRIPT

So, what became of the "species" that have haunted Eleazar Benavidez's sleep to this very day? Unlike the second set of bodies that would be flown out of Roswell to Fort Worth, Texas, two days hence on the B-29 *Straight Flush* (see Chapter 17), the trail becomes murky once Benavidez completes their delivery to the base hospital. We know that they were flown from Roswell to Wright Field the next day, after their overnight stay at the hospital, on a C-54 transport piloted by Captain Oliver "Pappy" Henderson. Henderson died in 1987, but not before telling his wife, family, and former crewmates from World War II about it. Henderson's wife, Sappho, and his daughter, Kathryn Groode, have given recorded testimony detailing what "Pappy" told them of his role in the Roswell event. Unfortunately, Henderson's crewmates from World War II were not his crewmates at Roswell in 1947, as we found when we interviewed each, to our disappointment. We have been able to locate only one of Henderson's crew who was on that July 8, 1947, "body flight" to Wright. Unfortunately for us, Technical Sergeant David Ackroyd had already passed away when we found him. When we asked his widow if her husband had ever discussed the flight with her, she replied, "All I remember about it is that one day during this thing he came home

The last remaining wing of the RAAF Base Hospital where the dead bodies and the live alien were taken from Hangar P-3 in a 1994 photo. The structure was taken down in the 1990s. (Photo courtesy of Tom Carey.)

early in a big rush and said, 'I need to get some things. I have to go on special flight with "Pappy" back East. I'll be home in a day or two.' That was all."[29]

Well, not quite. In 2008, we interviewed a former B-29 pilot from World War II, Joseph Toth, who was on the island of Tinian when the A-bombs were dropped. His niece who lives in New Mexico had called the International UFO Museum & Research Center in Roswell about him. (Such leads are forwarded to our investigation for follow-up.) At the time of the Roswell events, Toth was a captain in the Army Air Corps[30] flying B-29s out of Amarillo, Texas. He stressed that it was before the separation of the service into the U.S. Air Force in September of 1947. According to Toth, he was sent to "Wright and Patterson Field" (as he called it) to undergo a complete physical exam to determine if he could continue on flight status. During a lull while waiting for his physical, Toth wandered outside to have a look around. It was in the afternoon, as he recalled it, when a C-54 flew in and taxied "up real close" to the base medical facility and started to unload. He noticed something curious about this flight in that everyone associated with its unloading was moving very quickly, as though they were in a rush to complete the job for some reason.

It was then that he noticed men unloading three stretchers from the hold of the aircraft. On the stretchers were "three, short, grayish-bluish bodies with large heads." Viewing this scene from 150 feet away, Toth said that he could not see any details on the bodies other than the gross anatomy just described due to the distance and the fact that the men carrying the stretchers were moving so quickly. "That's all I saw." He remembered reading about Roswell in the newspapers at the time, and he especially remembered "a lot of the fellows at Wright talking about it after the C-54 unloaded." Toth failed his physical and ultimately left the service.[31]

15

"WHAT DID THEY LOOK LIKE, DADDY?"

They didn't look like they were from Texas.

—FRANK KAUFMANN, CBS 48 Hours, *1994*

T he late Len Stringfield, who was the first researcher to take UFO crash retrievals seriously, once met with a doctor he described as his "prime medical contact" for information regarding the alien bodies from Roswell. From that source he heard detailed descriptions of the unique body structure of the recovered crash victims.

Later, Stringfield was able to talk with a doctor who had participated in an autopsy of another such specimen. This new witness, according to Stringfield, provided a great deal of additional data: "From him, in time, I was able to envision the body entire," Stringfield wrote. "I learned of its internal chemistry and some of its organs—or, by human equation, the lack of them."[1] Stringfield, working with such information from his medical sources, was able to draw a number of conclusions: the being was humanoid, 3½ to 4 feet tall, and weighed about forty pounds.

The being's head was proportionally larger than a human head. The being had two large, round eyes, though one source did suggest the eyes were "Oriental or Mongoloid, deep-set and wide apart." Its nose was vague, with only a slight protuberance. Its mouth was a small slit and opened into a slight cavity. The mouth apparently did not "function as a means of communication or as an orifice for food ingestion," and there were no teeth. There were no earlobes or "protrusive flesh extending beyond apertures on each side of the head."[2]

The being's body had no hair on its head, though, according to Stringfield, one of his medical sources said that it was covered with a slight fuzz. The neck was thin, as was the torso. The arms were long and thin and the hands reached close to the knee. "A slight webbing effect between fingers was also noted by three observers," he wrote.[3]

"Skin description is not green," Stringfield continued. "Some claim it was beige, tan, brown, or . . . pinkish gray, and one said it looked almost 'bluish gray' under deep-freeze lights. The texture was described as 'scaly' or 'reptilian' and as 'stretchable, elastic, or mobile over smooth muscle.'" He noted that "under magnification, I was told, the tissue structure appears mesh-like. This information suggests the texture of the granular-skinned lizards, such as the iguana and chameleon."[4]

1978 composite drawing by the late Leonard Stringfield of a Roswell alien, based upon multiple discussions with a medical person who was present at the autopsies conducted at Wright-Patterson AFB. (Photo courtesy of Mrs. Dell Stringfield.)

Melvin E. Brown, a sergeant with the 509th Atomic Bomb Squadron at Roswell AAF in 1947, described bodies recovered at the crash scene when he was given the task of guarding a number of them after they were placed in the rear of a military ambulance truck. According to Brown, they were smaller than humans, and their skin was yellowish-orange in color and had a texture similar to that of a lizard—leathery and beaded, but not scaly. Given the circumstances of his opportunity to observe the bodies, the discrepancy between his and Stringfield's accounts of the skin color and texture could have been the result of decomposition or simply a difference in lighting conditions. Regardless, it is intriguing that Brown spoke of lizard-like skin, as Stringfield's sources did.[5]

Stringfield was also able to learn something about the internal organs and their structure. There was no apparent reproductive system and no genitalia. There was a colorless fluid prevalent throughout the body and, according to Stringfield's sources, "without red cells—there were no lymphocytes. Not a carrier of oxygen. No food or water intake is known. No digestive system or GI tract. No intestinal or alimentary canal or rectal area described."[6]

One secondhand witness's description of the bodies recovered at Roswell is suggestive of "insect-like" qualities. The late Frankie Dwyer Rowe was a twelve-year-old girl in 1947. Her father, now deceased, was a crew chief for the Roswell Fire Department at the time of the Roswell Incident. When the call came in to the fire station that there had been a crash north of town, Dan Dwyer and Lee Reeves were dispatched with the station's "tanker" (a pickup truck with a large, cylindrical water tank in the back) to the crash site. Arriving before the military could secure the site, Dwyer and Reeves got to see what had crashed. It wasn't an airplane at all, but an egg-shaped vessel of some sort that they did not recognize. And the bodies! Dwyer could see three diminutive, human-like-but-not-human creatures lying in the lee of the craft. With his focus riveted on the craft and the bodies on the ground, all of a sudden there was movement out of the corner of his eye. Walking in front of him, seemingly from out of nowhere, was a creature from another world! When Dwyer returned home that evening, he couldn't contain himself about what he had seen that day. When the question got around to, "Daddy, what did they look like?" his answer was a succinct, "Child of the Earth!"[7] With mouths agape, silence fell. To those of you not from the southwestern part of the United States, another name for the Child of the Earth is the "Jerusalem cricket." The former name is more popular with residents of the Southwest because of the morphology of this arthropod's head, which resembles in some ways that of a human infant. In our context, however, its head resembles strongly the descriptions of the head and facial features given throughout the years by alleged eyewitnesses to alien entities, including those of other witnesses regarding the aliens recovered from the Roswell crash. For us, the Child of the Earth image is undeniable in affirming the old saying that "a picture is worth a thousand words." Minus the antennae, you may indeed be looking at the face of one of the "victims of the wreck" from the Roswell UFO crash of 1947.[8]

Further confirmation of autopsies has come from another source: Dr. Lejeune Foster, a renowned expert and authority on human spinal-cord structure, who had a clinic in San Diego in 1947, was called on to perform a special assignment for the military. She was not sent to Wright Field, as one might assume, but rather to Washington, D.C., where she stayed for approximately one month.

Possibly because Dr. Foster had worked undercover for the FBI during the Second World War, and as a result had a top security clearance, she was flown to Washington to examine the spinal structures of the bodies retrieved near Roswell. She reported that her information from a high-ranking source was that one had been found alive but critically injured. According to the doctor, it had been rushed to Washington, where it soon died of its injuries.

Dr. Foster saw one or two of the bodies. Her task was to check their bone structure, spinal cords, and vertebrae, and to make comparisons between human anatomy and theirs. Because of her medical specialty, she was brought in to help answer the tremendous curiosity regarding the cervical spine, and the thoracic spine below and behind the chest that anchors the ribs. Dr. Foster was asked to examine the spinal canal that protects the spinal cord itself. Of special interest were the spinal nerves that carry impulses to move muscles or carry information to the brain from the sense organs in the skin, muscles, ligaments, and internal organs. Dr. Foster observed differences in the number of vertebrae, not so much with bone structure, but the absence of specific internal organs. She did not elaborate further.[9]

As did the other doctors who saw the bodies, Dr. Foster portrayed the beings as short, with proportionately larger heads than those of humans. She said they had "strange eyes."

According to family members, Dr. Foster was very upset upon her return from Washington. Her housekeeper, who was also a dear personal friend, sadly saw a change that only got worse. She became distant and seldom talked to anyone but her medical patients. As she had been debriefed, she had been told that if she talked about what she had seen she would lose her license to practice medicine and that she risked being killed! "Someone in the government is trying to keep me quiet," she often said.[10]

People always ask us, "Where are the bodies recovered from Roswell today?" In all honesty, we do not know for certain. What we *do* know is that, immediately following their retrieval, they were flown out of Roswell on two flights (July 8 and July 9) with an ultimate destination of Wright Field in Dayton, Ohio. Wright was a logical destination, because it was home to the Army Air Forces' Air Technical Intelligence Center (ATIC), which would later become the Foreign Technology Division (FTD) of the Air Force. Simply put, ATIC's mission was to examine, identify, and exploit "foreign objects" that came into its possession—the wreckage of a craft and crew emanating from off the planet, like that of a mundane Russian MiG 15, would certainly qualify as a "foreign object." Wright Field also had an aero medical facility better equipped to handle the bodies than either Roswell AAF or Fort Worth AAF. It appears that throughout the years, however, the bodies recovered at Roswell were "loaned around" to select, secure military facilities for examination. Curiously, we have two separate witness testimonies of seeing the bodies in Texas in 1964. One sighting was at Lackland AFB, and the other was at Randolph AFB. Both bases are located in San Antonio, and both bases have aero medical facilities (see Chapter 23). As for "the one that was alive," we have no first- or secondhand accounts that it accompanied its dead comrades on either of the two flights to Wright Field. Instead, we have stories that it went to White Sands near Alamogordo where captured German V-2 rockets were being tested, to Los Alamos or Sandia near Albuquerque where atomic research was being conducted, or that it never left Roswell alive. (See Chapter 11 and John Tilley's account concerning the rumors he heard about the one "live" alien having been shot by panicky sentries when Tilley was stationed at the Roswell base; see Chapter 23 and the reference to Jack Rodden Sr., who cryptically exclaimed several times to his son at the time of the incident, or soon thereafter, "They killed it!" Rodden Sr. did not elaborate to his son what he meant by that. Finally, there have also been stories from some throughout the years who claim to have seen the "live" Roswell alien at Wright Field (later Wright-Patterson AFB). One account has the alien being kept there until it died in 1952. Another account has someone running into it "in the basement" of one of the buildings there. More recently, a most credible account comes to us from the family of Marine Lieutenant Colonel Marion "Black Mac" Magruder (dec. 1997), whose Air War College class was flown to Wright AFB in the spring of 1948. While there, according to Magruder, the class was shown some of the Roswell UFO wreckage and the "live" alien (see Chapter 23 for the complete account). According to Magruder, he heard later that the alien he had seen was accidentally killed, year unknown, in an experiment being conducted on it. After the demise of the so-called "live-one," it appears that it joined its four dead Roswell comrades in their final resting place at the Dugway Proving Grounds in Utah, according to the former head of FTD at Wright-Patterson AFB, Colonel George Weinbrenner.

16

"WHO GOES THERE?"

As we have said, the 509th Bomb Group at Roswell Army Air Field in 1947, as the Air Force's first operational SAC unit, was perhaps *the* elite military unit in our country's entire Armed Forces. The RAAF was chosen to house the 509th Bomb Group because of its central location on the North American land mass, and it offered the added security feature of being situated on the ground "in the middle of nowhere." All of its officers and enlisted personnel were handpicked for its mission and had to pass extensive security background checks for the jobs they were intended to perform. Located just seven miles south of the city of Roswell itself, the RAAF was in a perpetual state of high security and readiness. Military police patrolled the fence perimeter that enclosed the base, as well as most of the sensitive areas within the base: the flight line with its compliment of Silverplates at the ready, the hangars and repair facilities, and the long, windowless, concrete building where the famed Norden bombsight was kept. The use of terminal force was authorized. (One witness we interviewed told of an airman who was shot and killed by an MP near the flight line as he was running toward his aircraft; he was apparently late for takeoff and attempted a shortcut across an area where the use of lethal force was authorized, and failed to stop when challenged.) At the main gate located at the north end of the base, facing the city, armed MPs controlled the entrance from an enclosed security booth. In addition to an intelligence office, the base also had a counterintelligence office, whose civilian-clad officers checked out possible security risks and breaches, trying to combat the very real threat of espionage. Airmen stationed at the base in those days have all told us of the tight security that was always in place, twenty-four hours a day, every day. The need for security was "drummed into us" from the moment of their first arrival, as one witness told us. "Need-to-know" has always been a way of life in the military, but it was carried to an extreme at Roswell, as 509th personnel were told not to discuss with anyone, not even their wives and family, at the risk of federal imprisonment, anything that took place on the base or in the course of carrying out their duties.

As a reminder, just in case anyone might have thought that the "loose lips sink ships" mentality of WWII was a thing of the past, large signs were placed at strategic locations on the base that warned:

WHAT YOU HEAR HERE

WHAT YOU SEE HERE

WHEN YOU LEAVE HERE

LET IT STAY HERE!!

So it is that we have spoken to many widows of former 509th airmen who were at Roswell in 1947 who have told us, when asked about their husbands' knowledge of the Roswell Incident, "He never said anything about it." Although we find it difficult to understand today how someone could remain silent for the rest of his life, not even telling his wife that he was witness to the events that are the subject of this book, we respect that possibility whenever we encounter it. We admit that we couldn't do it, but there is no doubt that more than a few 509th personnel from America's "greatest generation"—whether out of a sense of patriotism or fear of retaliation— did just that. On the other hand, what easier way is there for an elderly widow to terminate a conversation with a complete stranger on a subject that she may not want to discuss—for whatever reason—than to say that she knows nothing because her husband told her nothing? End of conversation. Fade to black.

Fortunately for us and for history, our witness list of those who have talked to us about the 1947 Roswell events now totals more than 600 and counting.[1] This chapter will demonstrate to the reader how we fill in detail and flesh out the timeline in our story. It pertains to the preparations for the alleged July 9, 1947, "body flight" to Fort Worth, Texas.

This was the unscheduled, special flight of a single B-29 known as the *Straight Flush* from the 393rd Bomb Squadron that flew from Roswell Army Air Field to Fort Worth Army Air Field at the suspiciously low altitude of only 8,000 feet (B-29s normally flew at an altitude of about 25,000 feet). In place of its normal payload that day was a cargo consisting of a single 4-foot by 5-foot by 15-foot wooden crate in the bomb bay. Surrounding the crate in the bomb bay was a contingent consisting of a half a dozen or so armed security guards. Because the bomb bay of a B-29 was not pressurized, the *Straight Flush* had to fly low enough so that the security guards in it would not succumb to oxygen starvation. The low altitude also suggests that there might have been something biological in the crate. Upon landing in Fort Worth, the bombardier on the flight was overheard to remark that he recognized an old friend—a mortician—from his college days among the officers waiting in the greeting party on the tarmac. He was immediately ordered to "Shut up!" by the aircraft commander. The enlisted crewmembers on the flight were by this time starting to put two and two together. They had heard the talk back in Roswell about a "spaceship" crash north of town a few days earlier. The day before, they read in the *Roswell Daily Record* and heard on the radio that a downed flying saucer had been recovered by the 509th and was being shipped to "higher headquarters"—that meant Fort Worth. There had also been talk of strange little bodies. And when the wooden crate was loaded into the bomb bay of their B-29 under armed guard, all of the enlisted crewmembers were ordered to stand at the far wingtip and face away from the aircraft, so they could not see what was going on. This order, of course, only served to embolden some of the crewmembers to take a glance. On the flight to Fort Worth, therefore, it did not require much of a stretch to think that the large crate in the bomb bay possibly had something to do with the "spaceship crash" that everybody was talking about back home (the officers on the flight already knew, as they had been briefed separately prior to takeoff). Now, after hearing about a mortician waiting for their flight, and then hearing one of the officers on their flight boast, upon taking off for the flight back to Roswell, that they "just made history," the crew started entertaining the thought that the wooden crate might have had something to do with the "little bodies." One of the crewmembers was especially convinced of this. Unknown to the others at the time, Private

Lloyd Thompson, a gunner on the *Straight Flush* that day, recognized a flight physician from the Roswell base hospital who had treated him a week earlier for a sore shoulder, and who was now on the flight to Fort Worth with the security detail attending to the wooden crate. On the hour-long return flight to Roswell, after the wooden crate had been unloaded in Fort Worth, the *Straight Flush* flew at its normal altitude of 25,000 feet.

The information we have been able to gather about the flight of the *Straight Flush* is the result of interviews we conducted with three of its crewmembers from that July 9, 1947, flight: Private Lloyd J. Thompson, a waist gunner and assistant flight engineer who passed away in 2003; Staff Sergeant Arthur J. Osepchook, another gunner; and Staff Sergeant Robert A. Slusher, the radioman. All other crewmembers that we had identified as being on that flight either have passed away, were killed, or refused to talk to us when we came calling. Besides Osepchook and Slusher, there was only one other member of the crew known to possibly have been alive: Lieutenant James W. Eubanks, the flight's navigator.

Our investigation located James Eubanks in 2003, but after learning the subject of our telephone call, he quickly terminated the conversation with, "I can't remember," and hung up. He soon changed his phone number to unlisted and became unavailable to us. Eubanks passed away in 2005.

Lloyd Thompson was the first former crewmember to come forward in the late 1980s when he wrote a letter to a UFO group, Just Cause, who then passed along his letter to UFO crash/retrieval researcher, the late Leonard Stringfield (1920–1994). Beginning in the late 1970s, Stringfield had published a series of unsourced monographs about alleged UFO crashes, a subject that was taboo for UFO researchers up to that time because of its sensationalist nature (Thompson's account, authored by Stringfield, appeared in the *MUFON UFO Journal* in 1989). However, Thompson would not allow Stringfield to identify him by name, thereby forcing Stringfield to refer to him only as "Tim" in the article. It wasn't until 2000 that Thompson "came out of the closet" when he gave a public presentation about the *Straight Flush*'s flight to a local UFO group in the town where his son lived. Schmitt (as well as Stringfield) had interviewed Thompson several times throughout the years, but it was during a 2005 interview that Carey had with Thompson's son Lowell, after his father had passed away, that we were able to learn additional details about the flight and its aftermath that his father never revealed publicly.

Robert Slusher (now deceased) had always been open and up-front with us regarding what he knows and what he doesn't know about the flight, without embellishing. In 2002, he appeared on-site at the old Roswell base to tell his story for our two-hour Sci Fi Channel special, *The Roswell Crash: Startling New Evidence*. We interviewed Arthur Osepchook in 2004. He was very cautious about the flight and could offer no additional details of his own on the subject beyond noting the unusual nature of the flight and affirming the accounts of his crewmates, Thompson and Slusher. He was expansive, however, in recalling the excitement and talk on the Roswell base at the time about a "spaceship crash" and the "little bodies" supposedly found among the wreckage. Osepchook passed away in 2015.

Several years ago, a woman who operated a local museum in another state visited the International UFO Museum & Research Center in Roswell. She claimed to know a Native American "back home" who had been stationed at Roswell in 1947 and had "guarded the bodies."[2] She said that he was still alive, and she left his name with the museum director. She warned, however, not to try to contact him directly, as he would not talk to anyone who had not been introduced through her. After several years of trying without success to arrange a telephone or an in-person interview with the gentleman by going through the woman, we decided to take matters into our own hands by calling him directly. We felt that we had nothing to lose at that point, in 2005, as the years were rapidly slipping away. Edward Harrison answered the telephone himself, and, contrary to the woman's warning, agreed to talk, however cautiously, about his participation in the 1947 events at Roswell.

Corporal Edward Harrison was the corporal of the guard in charge of a small detachment of Native Americans at the RAAF ordered to report to the far southwest corner of the base for a special duty assignment. He could remember the name of only one of the airmen in his charge that day, "a full-blooded Omaha" by the name of PFC James J. Lyons (now deceased). He did not remember the exact date, only that it occurred in July of 1947, "during the time of all the talk on the base about the crash of a spaceship" (we believe the date was July 8). When Corporal Harrison and his men reported to the designated location at the far end of the base as ordered, they were somewhat surprised to be met by their commanding officer, Lieutenant Colonel John S. Loomis of the 1027th Air Material Squadron, who was sitting in a jeep in front of the locked gate of an 8-foot-high chain-link fence enclosing a large tent. Harrison had never known a tent to be in that location on the base before—near the trash incinerator—and, judging by the fresh earth piled up at the base of the fence posts, he surmised that the fence and tent had only recently been erected. He estimated the perimeter dimensions of the fence to be approximately 20 feet by 20 feet. The tent inside was about 15 feet by 15 feet by 10 feet at its central peak. It was one of those old olive drab Army tents used in WWII to sleep four to six soldiers (uncomfortably) in the field. After exchanging salutes with Colonel Loomis, Harrison was ordered to post armed guards around the outside of the fence with orders to "shoot anything that isn't a rabbit." According to Harrison, this was an unusual but not unheard-of order, considering the nature of the assignments his detachment had been given in the past. What was unusual to Harrison, however, was the sudden appearance of a tent and fence in the remotest part of the base, which he was being asked to secure with terminal force, if necessary. "Something not normal is going on here," Harrison thought to himself, as he saluted his understanding and acceptance of the order to Colonel Loomis, who then quickly departed. After instructing his men, Harrison departed by jeep as well. He had arranged for several changes of the guard covering the next twenty-four hours. Early the following morning (July 9) Harrison drove his men to the tent to relieve the guards on duty, only to discover that the guards, the fence, the tent, and whatever was in it were gone! Scratching his head in bewilderment at this latest surprise, Harrison noticed a freshly graded dirt road and a set of large tire tracks in the soft morning earth that came from the general vicinity of the flight line up to where the tent had been, and then continued off in the direction of Bomb Pit #1—another highly secure location on the base, where the atomic bombs were loaded into the B-29s. When asked by our investigation almost sixty years later if there was a foul smell coming from the tent, Harrison replied, "Maybe, but it was hard to tell because of the ever-present smell of aviation fuel from the aircraft engines that permeated the air in that part of the base." Not having further orders or a need to know what was going on, Harrison and his men departed the site and never returned.[3]

Other men were able to get considerably closer to Bomb Pit #1. Corporal William L. Quigley had already finished working his day shift in the armaments shop of the 393rd Bomb Squadron[4] at the RAAF, and the evening found him in his room in the enlisted men's barracks getting ready to turn in, when the sergeant of the guard unexpectedly appeared at his door, saying, "Corporal Quigley, I have a job for you. Follow me!" Not particularly pleased at this turn of events, as it was getting late and he had already worked his eight-hour shift that day, Quigley unhappily complied. Both men then went outside and got into a waiting jeep that quickly whisked them across the flight line to Bomb Pit #1. Quigley knew that this was one of the most sensitive locations on the base, but he didn't know what this had to do with him. He was ordered to get out of the jeep and to stand by.

"The Pit," as it was known, was a large, cement-covered, rectangular hole in the ground containing a hydraulic lift big enough to raise an A-bomb 10 feet up and into the bomb bay of a waiting B-29. When the bomb-loading procedure was in progress, armed MPs normally secured the area, and a 7½-foot-high canvas screen-type barrier was placed around the perimeter of the bomb pit directly underneath the B-29 so that no one on the outside could see what was going on inside.

Quigley was handed an M-1 carbine rifle and joined three others, two officers and one enlisted man, who were also holding carbines and who were also obviously standing by with that "what am I doing here?" look on their faces. It was only now that he became aware of a B-29 not far from where they were standing, partially obscured by canvas screens. (In addition to the canvas screen barriers in place around The Pit proper, this time there was a second set erected in parallel around the inner ring of screens that, according to Quigley, formed an alley where the guards were to walk their security details.) The four were then led over to the B-29 and taken inside the outer ring of canvas screens and were given their orders. "I felt like I was walking in a tunnel," Quigley would say in 2006. "I couldn't see in, and I couldn't see out."

Each of the four guards walked a quarter of the perimeter in this circular tunnel, as assigned. Quigley's station was to walk back and forth from the tail of the aircraft to the right wingtip (he remembers seeing the tail number of the B-29 he was guarding that night—#291). His orders were to challenge anyone and anything that moved, and the use of terminal force was authorized. A few hours into his shift, around 3 a.m., Quigley thought he heard something outside the perimeter. "Who goes there?" he challenged. Nothing. "Stop and identify yourself!" Still nothing. Knowing that his orders were not to leave the "tunnel" under any circumstances, Quigley continued walking his detail. A little later, he noticed some movement against one of the outer screens. This time instead of challenging, "I just took my rifle butt and rammed it against the screen as hard as I could where the movement was and called for the sergeant of the guard, but no one came. I don't know what I hit, but it was something solid. I heard later that I had butt-ended a 'CIA-type' [using today's terminology] who was checking security around The Pit that night." At the end of Quigley's shift, instead of heading back to his barracks for some much-needed sleep, he and the other three guards were driven over to squadron headquarters, where all were told to sign a piece of paper that stated, as Quigley tells it today, "It didn't happen" (meaning that Quigley did not guard The Pit that night/morning). Quigley thought this to be highly unusual at the time, but because he was more concerned about getting some sleep, he quickly signed the statement and left. It would be the only time he was ever asked—actually, ordered—to sign such a statement.

Recalling the episode, Quigley prefaced what he was about to say with, "I have never talked to anyone about this before. There have been phone calls, but I always told them that I didn't know anything. I understand, however, that the ban on talking about this has been lifted, and I can now talk. You will be the first to hear this." He continued, "I didn't know what to make out of it at the time, but it was a highly unusual situation. The double row of canvas barriers was not normal and way, way out of the ordinary." This suggests that the second row of canvas screens was erected in this instance to prevent the guards on duty from seeing anything that was going on around them. Quigley went on:

> *I had never pulled guard duty in my life, either before or after that night. Normally, the MPs provided the security for an operation like that, but I heard a little later that something "bigger than the A-bomb" was going on, and that the MPs didn't have a high enough security clearance for it. That would explain why I was chosen for this special duty, because I possessed the highest security clearance that an enlisted man could obtain—higher, at least, than anyone else I knew. I guess they could find only one other enlisted man in my squadron with the required clearance and therefore had to substitute two officers to fill the other two guard spots. This took place during the time of all the buzz on the base about UFOs. All the time I was walking my detail that night, I never heard any sound whatsoever coming from The Pit area behind the inner ring of canvas screens. I believed at the time that I was guarding a B-29 with an A-bomb aboard in its bomb bay. A few days later, after all the excitement had died down, I heard that I had guarded the alien bodies that everyone was talking about that were stored in The Pit that night in preparation for flying them to Texas the next day.*

We already knew from our investigation that the *Straight Flush* was the B-29 that allegedly flew several alien bodies to Fort Worth, Texas, on July 9, 1947 (several other bodies, we believe, had been flown out a day earlier, directly from Roswell to Wright Field in Dayton, Ohio, on a C-54 piloted by Captain Oliver "Pappy" Henderson). Because we have not committed to memory the tail numbers of every B-29 in the 509th Bomb Group's inventory, a quick check was made by Carey for the tail number of the *Straight Flush*. Fully expecting that it would turn out to be #291, it was a shock and a disappointment to discover that it was in fact #301. This would seem to suggest that William Quigley's interesting story had nothing to do with the Roswell Incident—that it was just another situation involving high security on a base full of high-security situations.

Out of curiosity, however, and to put a period on this story, we wanted to know which B-29 bore the tail number #291. Further research showed that #291 belonged to a B-29 known as the *Necessary Evil*.[5] That name rang a bell with Carey, who recalled that he had a copy of a picture of a B-29 with that name that had been sent to him by Robert Slusher. Sure enough, Carey located it, and there was Robert Slusher standing right by the *Necessary Evil* in clear view with the rest of its crew! Thoroughly confused now, because Slusher was known to have been a crewmember on the July 9 flight of the *Straight Flush*, Carey decided to call Slusher for an explanation of this apparent contradiction.

The articulate Bob Slusher was up to the challenge. "The *Necessary Evil* was our regular aircraft," Slusher told Carey, "and we were surprised when we were told to report to another aircraft, the *Straight Flush*, for a special flight to Fort Worth that day (July 9, 1947). We found out later that the cargo we were carrying that day—a large wooden crate—would not fit properly in the forward bomb bay of the *Necessary Evil*. So at the last minute, the *Straight Flush* was substituted in its place."[6]

The B-29 *Necessary Evil* [#291] was originally scheduled to fly the large wooden crate to Fort Worth AAF, but it was too big to fit into the plane's forward bomb-bay compartment. (Photo courtesy of the U.S. Air Force.)

We can now meld these separate witness accounts into a scenario that suggests the impromptu security measures that were employed to keep the secret during the twenty-four-hour period preceding the "body flight" from Roswell to Fort Worth on July 9, 1947. This flight involved a *second* set of bodies brought to the base. Unlike the previous set of alien bodies—found closer to Roswell in Chaves County and brought in two days before—these had been found closer to Corona in Lincoln County, and all were dead. They also smelled. After an aborted attempt at conducting an autopsy at the base hospital (terminated due to the smell), it was decided to fly the bodies out as soon as possible to a facility that was better equipped to handle such a procedure. In the meantime, the bodies were placed in body bags, packed in dry ice within a large wooden crate, and removed to the remotest part of the base in the hope that the smell, if it could not be contained, would at least not reach the rest of the base. To block its view from curious eyes, the crate was placed in a tent, and to prevent whoever was guarding the tent from taking a peek inside, a tall fence was erected around it. Enter Corporal Harrison's high-security detail at the tent with shoot-to-kill orders, but with no idea what it was they were guarding. When it came time to move the crate over to Bomb Pit #1, it was done under cover of darkness in order to minimize the chance of being observed (sometime after midnight on the morning of July 9), and the twin military doctrines of "compartmentalization" and "need-to-know" were employed to keep Harrison and his men from learning too much about what might be going on. Therefore, instead of accompanying the crate over to the bomb pit, they were dismissed just prior to its transfer.

After the crate was lowered into The Pit, the *Necessary Evil* taxied over from the flight line and assumed "loading position" over the bomb pit for the night (the actual loading of the crate into its bomb bay was scheduled for the following morning—in daylight). Who, then, should and could guard the crate? They couldn't use any of Harrison's men, because word would surely get back regarding the new location of the crate, and they might then be able to connect the dots. They also did not want to chance a security breach by expanding this responsibility to any other unit on the base, such as the 1395th MP Company (most MPs also did not possess high enough security clearances for this situation). It was decided that the best way to prevent leakage was to keep everything compartmentalized within the 393rd Bomb Squadron. A quick roster check of the men of the 393rd Bomb Squadron revealed only two available enlistees (one of whom was Corporal William L. Quigley) with security clearances high enough to permit them to stand guard for something that was "bigger than the A-bomb."

The testimonies of Edward Harrison and William Quigley, neither of which has ever before been published, serves to put an exclamation point on the testimony of the crewmembers who were aboard the strange flight of the *Straight Flush* on July 9, 1947, that something highly classified and unusual had taken place. Taken in isolation, each participant's testimony could be dismissed as nothing more than just another day in the life of someone stationed at an SAC base in the late 1940s. But our job is to connect the dots, to put the pieces of the puzzle together. And we have sufficient dots and pieces of the puzzle in the form of related and credible testimony to do that. When we plug in Harrison's and Quigley's accounts of the strange and secret goings-on at the fenced-in tent and Bomb Pit #1 to what we know about the unusual and highly classified flight of the *Straight Flush* the following day, a more complete and fortified picture of the Roswell time line comes into focus, as well as some of the steps employed by the military to contain knowledge of the unfolding events.

17

"BOYS, WE JUST MADE HISTORY!"

The entry in the *Straight Flush*'s flight log for July 9, 1947, read, "DEH, Ship #7301. B-29. Cross-country. Forth Worth and Return. Flight time: 1 hr. 55 mins."[1] It was the day after Brigadier General Roger Ramey had put an end to all the excitement and public clamor for more information about the flying saucer reported to have been recovered by the 509th Bomb Group at Roswell Army Air Field.

It was shortly after lunch when a flight crew went to the skeet range to test their shooting skills. They had spent the morning attending "ground-training classes" in Russian history, Russian language, and hand-to-hand combat, and their official duties for the day were completed. The aircraft commander, Captain Frederick Ewing, had shattered forty-eight of fifty targets. Private Lloyd Thompson, who was on the base skeet team, did nearly as well with forty-seven out of fifty, while Staff Sergeant Robert Slusher, a radioman, left to join baseball practice for the base team, the RAAF Bombers. Still, the buzz about the flying disc and the "little bodies" preoccupied the shooters, and the NCO in charge of the skeet range talked about nothing else. It was about 3 p.m. when the officer of the day (a rotating guard position) came by the skeet range in a jeep and told the men to stay together and stand by. "This could be serious," said Captain Ewing. Slusher was pulled from the baseball field and rejoined the rest. Something was up. A bus then picked them up and transported them to the flight operations building near the flight line.

All the other flight crews had been released for the day. The operations officer, Major Edgar R. Skelley, instructed the crew to stand by, because he had an unscheduled flight for them. Not knowing what to expect, the crew thought that it might be a last-minute test flight to prep a plane for a mission the next day. This thought did not change when the enlisted crewmembers were ordered to preflight a B-29, nicknamed the *Straight Flush* (#301), that was waiting on the concrete tarmac just behind the flight operations building. This was because the *Straight Flush* was not this crew's regular aircraft. Theirs was a B-29 nicknamed the *Necessary Evil* (#291). The preflight of a B-29 was an involved, lengthy affair because it had four separate engines, each with individual systems, that had to be checked out. The entire process took about an hour to complete. During this time, the officers of the crew had remained inside the flight ops building, where they were cleared and briefed for the impending flight. With the preflight completed, the commanding officer of the 393rd Bomb Squadron, Lieutenant Colonel Virgil Cloyd, appeared and told the crew, "This is a routine mission. Do exactly as you are told, and don't discuss it." Major Skelley then instructed the crew to board the *Straight Flush* and taxi it over to Bomb Pit #1.

Bomb Pit #1, known to flight crews simply as The Pit, was housed in a cul-de-sac located to the southwest just off the main tarmac of the flight line. Signs were posted around the bomb pit area that read:

LETHAL FORCE WILL BE
USED ON ANYONE
TRESPASSING WITHIN
200 FEET OF THIS AREA

The only areas of the base that had unobstructed views of the bomb pit were the air traffic control tower near the flight ops building and portions of the flight line. While the huge airplane positioned its open, forward bomb bay directly over the pit for loading, it was then that the crew noticed that The Pit was covered by a large, canvas tarp. But there was no atomic bomb in the bomb pit that afternoon. "Did you see what I saw?" Corporal Thaddeus Love, a tail gunner who was on the flight, was overheard asking another crewmember. "No talking!" was Captain Ewing's quick retort.

The enlisted crewmembers and NCOs were ordered to deplane and stand at the far wingtip of the aircraft, facing away from the bomb pit. As this was unusual, even for A-bomb loading, they still did not know what was going on. They also did not know that their own aircraft, the *Necessary Evil,* had spent the night and part of that day parked over The Pit while a ground crew tried in vain to ready it for this flight. When it still was not ready by midafternoon, it was decided to substitute another aircraft, the *Straight Flush,* in its place. The canvas cover over the bomb pit

The B-29 *Straight Flush* [#301] was substituted for the *Necessary Evil* to fly the alien-containing large wooden crate to Fort Worth AAF on July 9, 1947. (Photot courtesy of the U.S. Air Force.)

was pulled away, exposing a large, rectangular wooden crate. All indications suggested it had been hastily constructed, as it was unpainted and unmarked. Its size—approximately 4 feet high, 5 feet wide, and 15 feet long—made for a snug fit as it was hoisted into the forward bomb bay of the waiting B-29. Four special security guards dressed in full MP Class-A uniforms positioned themselves at each corner of the sealed box. One of them was a major, one was a captain, one was a lieutenant, and one was an NCO. Three additional MPs positioned themselves inside the aircraft in the fore and aft crew compartments. After the bomb bay doors closed, the enlisted men and NCOs had to wait before reboarding the aircraft, while Lieutenant Martucci, the bombardier on the flight, conducted a security check of the cargo in the bomb bay. It wasn't until the *Straight Flush* was airborne that the rest of the crew were told of their official destination: Fort Worth Army Air Field, home and headquarters of the "Mighty Eighth" Air Force, to which Roswell's 509th Bomb Group was attached. This was definitely no quickie test flight.

The normal flight time from Roswell to Forth Worth was about one hour, but with the aid of slight tail wind, the flight this day took only fifty-five minutes. Upon landing at the base, they taxied the *Straight Flush* across the tarmac over to a waiting contingent of officers, a development that only added to the mystery of the unidentified cargo in the bomb bay. Lieutenant Martucci was overheard to remark excitedly to no one in particular that he recognized one of the officers in the greeting party as an old friend from his college days. "Hey! I know that guy. I went to school with him. He's a mortician." This implied that whatever was in the crate might be related to his friend's profession. Lieutenant Martucci's reminiscence was cut short with a stern admonition from Captain Ewing to "Shut up!" that came through the crew's earphones loud and clear.

Once the aircraft came to a stop, the crate was unloaded. One of the crew wasn't sure, but he thought he remembered seeing a gurney being wheeled out toward the crate. The enlistees and NCOs in the crew were ordered to remain on board, while all of the officers disembarked the aircraft to talk to the officers in the receiving line. Lieutenant Martucci deboarded the aircraft with the MPs in the bomb bay and immediately went over to chat up his old friend. The entire party plus the crate and the gurney then departed for one of the gray buildings lining the flight line, after which the remaining crewmen, who had been waiting in silence for fifteen minutes, were permitted to leave the aircraft to stretch their legs. They had been ordered, however, not to talk to anyone on the ground. Sandwiches and soda were brought out to them for a snack, which, considering the time of day, turned out to be their dinner. When they were just about finished eating, the flight officers, as well as one or two of the MPs, returned to the aircraft. Captain Ewing and Lieutenant Martucci, however, were not with them. Just then, Ewing and Martucci could be seen exiting the flight operations building with another officer. All three walked quickly across the tarmac and boarded the *Straight Flush*. Major Jesse A. Marcel, the intelligence officer of the 509th Bomb Group in Roswell, who had been flown to Fort Worth the day before with a planeload of wreckage that he had collected from the crash site (and who had been unwittingly used as a prop by General Ramey at the press conference the day before), was a "guest passenger" on the flight back to Roswell. One of the crewmembers observed that Marcel "appeared to be down and very sad that day." After everyone had boarded the aircraft and had settled in for the return flight, the aircraft commander, Captain Ewing, who had ordered Lieutenant Martucci to keep quiet moments before, now issued another curious but stern lecture aimed at the entire crew. They were told in no uncertain terms that this had been a "routine flight," and that they were never to say another word about it—even to their wives! "Forget everything that you saw today!" One crewman would later remark, "For weeks, rumors were plentiful, and we were hard-pressed to maintain the silence we had been ordered to keep." After the *Straight Flush* lifted off the runway at Fort Worth for the return flight to Roswell, according to surviving crewmembers, the irrepressible Lieutenant Martucci apparently could not contain himself any longer and was heard to exclaim to everyone within earshot, "Boys, we just made history!"[2]

This account of the flight of the *Straight Flush* suggests that something more important than General Ramey's laundry was being flown to Fort Worth that day. The secrecy of an unscheduled flight, the security surrounding an unmarked wooden crate, a low-altitude, cross-county flight, the curious behavior of certain officers on the flight, a flight surgeon dressed as a security officer, a mortician waiting for the flight, the gurney, the stern admonitions to keep quiet, and the remark about making the history books were all out of the ordinary—even if A-bombs were known to be aboard. When interviewed many years later with the benefit of hindsight and reflection, the surviving crewmembers to a man stated that they definitely suspected on the return flight, and believe now, that several alien cadavers were inside the wooden crate they had flown from Roswell to Fort Worth that day. In answer to the question regarding the possibility of a foul smell emanating from the crate, all stated that they were unaware of such a smell, but they were not permitted anywhere near the crate.

The flight crew that day, from the 393rd Bomb Squadron at the RAAF, was led by Captain Frederick Ewing, the aircraft's commander and pilot. Ewing was killed in an unfortunate air accident five years later while piloting his B-47 Stratojet medium bomber in an attempt to help another B-47 that was having difficulty with its landing gear. Ewing's copilot on the *Straight Flush*, Lieutenant Edgar Izard, retired from the Air Force and was last seen selling insurance in Roswell in the 1950s. To our knowledge, he was never located and interviewed. His wife was located in 2008 and confirmed his passing some years prior. She claimed to know nothing of the Roswell events of 1947. Lieutenant Elmer Landry was a fill-in on the July 9 flight for Sergeant David Tyner as the flight engineer. He too has never been interviewed and is presumed dead. Lieutenant James W. Eubanks, the flight's navigator, was located in Forth Worth. In a 2003 telephone interview with Carey, when informed of the nature of the call, Eubanks simply said, "I can't remember," and hung up. Follow-up telephone calls by Carey in the ensuing months went unanswered. Eubanks then changed his phone number to an unlisted number. It is now known that Eubanks passed away in 2005. Lieutenant Felix Martucci, the talkative bombardier and security officer on the flight who made the "Boys, we made history!" statement, was located in 1990 living in a military retirement community in San Antonio, Texas. Because of his suggestive statement on the flight, our investigation especially wanted to interview him to find out what he had meant by it. We also wanted to learn more about his mortician friend, and why he was waiting for his flight that day. Schmitt had briefly spoken to an unresponsive Martucci on the telephone, and when Len Stringfield, from whom we had initially learned about Martucci, followed up with a call to Martucci on his own, a woman answered and shouted a loud "No!" into Stringfield's ear before slamming down the phone. A hoped-for, in-person, "ambush" interview with Martucci at his residence in San Antonio unfortunately fell through when Martucci failed to appear. He subsequently obtained an unlisted telephone number to thwart efforts to contact him, thereby avoiding having to answer the questions. Martucci passed away in 1996—without ever revealing to anyone the true details of his self-proclaimed history-making flight.

Of the enlisted men and NCOs on the flight, Corporal Thaddeus Love died six months before we located his widow in the early 1990s with the help of Barbara Dugger and Christine Tulk, the granddaughters of the late Chaves County sheriff, George Wilcox, who was heavily involved in the Roswell Incident. Both were also schoolteachers, one of whom (Tulk) by coincidence worked at the same school as Love's widow. After informally learning of their mutual connection to Roswell at school, Tulk told Dugger, who also spoke to the woman. Love's widow did not deny anything that Tulk and Dugger were saying and suggesting, but she did not say much on her own either. Dugger then called Schmitt, whom she already knew from an earlier interview, and it was quickly decided that an in-person interview was in order. Schmitt, Dugger, and Tulk then drove from Roswell to El Paso. Schmitt got out of the car and went to the front door of the widow's house, while Dugger and Tulk remained in the car. Schmitt knocked on the door. The door opened just a crack so that Schmitt could barely see the woman's face. After introducing himself, Love's widow responded to Schmitt, "Yes, I know who you are. But, before my husband died, he asked me never to say another word about this, and I have no intention of breaking that promise. I wish I could help you. I'm very sorry." With that, she closed the door, and that was that. What is the explanation here? Never say anything about a *weather balloon?*

Robert Slusher introduced himself to our investigation in 1990 after a talk by Schmitt and fellow Roswell investigator Kevin Randle at the Alamogordo Civic Center. He was able to corroborate much of what Lloyd Thompson had told Stringfield and Schmitt previously about the July 9 flight. He was sincere, did not embellish, and what he did not know, he simply stated that he did not know. He kept his testimony to what he personally knew, and his story has not changed throughout the years in our follow-up interviews. Slusher was among those Roswell witnesses who were presciently brought to Washington, D.C., in 1991 by the Fund of UFO Research for the purpose of videotaping their individual testimonies for posterity in *Recollections of Roswell.* Slusher also appeared on location at the site of The Pit on the extinct base in Roswell to tell his story for our 2002 Sci Fi Channel production, *The Roswell Crash: Startling New Evidence.*

Arthur Osepchook was interviewed by Carey in 2003 and 2006. Although Osepchook could not recall many of the details, he remembered "the crate flight" and the feeling at the time that "something big was going on." Carey went through the list of crewmembers on the July 9 flight, all of whom Osepchook was able to confirm as being on that flight. He confirmed the trip to the skeet range and especially the air of excitement on the base about the rumored crash of a "spaceship" with "little bodies" inside. Even more than his colleagues, Osepchook was emphatic that he knew there was *something*—"something special"—in the crate. He did not elaborate on how he knew this or what he thought might have specifically been in the crate, but his recollection of postflight reaction no doubt contributed to his notion that something highly unusual was afoot, even before the *Straight Flush* touched down in Roswell. "We were told that 'nothing was going on' in order to kill such speculation on our part," he said. "Then, after we landed, a big meeting was called of the entire squadron in one of the hangars to tell us that there were no such things as *flying saucers* and that [a crash of one] didn't happen." Osepchook told Carey that he had "papers" from his military days that were stored away somewhere, and he promised to look for them.

Lloyd Thompson began corresponding with the late Cincinnati UFO researcher Leonard Stringfield in 1989. A responsible, "by the book" member of America's "greatest generation," Thompson felt compelled to get the truth out about the Roswell Incident, but wished to remain anonymous in doing so. It was for this reason that Stringfield would refer to Thompson only as "Tim" in his sixth published monograph about alleged UFO crash/retrievals as well as in an article published in the November 1989 *MUFON UFO Journal* ("Roswell & the X-15: UFO Basics"). A cautious and circumspect man, Thompson was invited but chose not to attend the 1991 *Recollections of Roswell* assemblage of Roswell Incident witnesses in Washington, D.C. In 2000, however, Lloyd Thompson finally went public when he gave a presentation about his participation in the Roswell Incident to a local UFO group in the town where his son lived. It is from Thompson's account in the aforementioned MUFON publication regarding the July 9, 1947, flight, from his 2000 presentation, as well as from Schmitt's own telephone interviews with him, that we have drawn to construct the basic history of that flight. Thompson passed away on April 13, 2004.

In conducting an investigation such as this, we have found that surviving family members will sometimes talk when the departed eyewitness would not—not always, but sometimes. Further, family members can also supply additional details to information already supplied by a deceased witness. Such was the case with Lloyd Thompson. Carey called Thompson's son Lowell in 2005 after he learned of Lloyd's death from his widow. Lowell Thompson was able to provide additional background information about his father that was of a personal nature, as well as new details concerning his father's involvement in the Roswell Incident. According to Lowell, his father never said a word about Roswell until 1984, after reading the first book about Roswell. After dinner one evening, he asked his family to return to the dinner table.

He was in a somber mood. I thought that he was going to tell us that he was divorcing my mother! He held up a copy of *The Roswell Incident* and said, "I didn't think that I would ever be able to tell you this. I thought it would be secret until my death. But I checked this book out from the library last week. I've gone through it, and they pretty much have everything in here." He went through the basic story and then showed us an old logbook of his with an entry for a flight which, I believe, was dated July 9, 1947. So, that's how it all got started. Later, he got in touch with Stringfield.

This postprandial revelation also explained for Lowell why his father had always encouraged him to take out UFO books when they went to the library. "Ever since I was eight years old, wherever we lived, he told me to do this, but never explained why. My father read a lot of UFO books. He also had me clip from the newspaper any articles or news stories about UFOs and put them in a scrapbook. Now I know why."

From the moment the wreckage from an unknown aerial device and the remains of its crew started arriving at the RAAF, until arrival at their final destination at Wright AAF in Dayton, this chapter has illustrated the classic, historical, military reaction to a UFO event. Secrecy, "compartmentalization," "suspension of disbelief," intimidation, and so on were all part of the extraordinary security measures employed. Hasty and momentous decisions, however seemingly myopic to us now, were made perhaps out of panic, perhaps out of ignorance, all for the primary purpose of keeping the secret—not only from the Russians, but also from the American people, as we will see in more detail in the next chapter.

18

"IF YOU SAY ANYTHING, YOU WILL BE KILLED!"

Today, we would refer to the threats made by some members of the military as civil rights violations, or even war crimes if committed in time of war, and their perpetrators would be brought to a swift justice and punished severely. Especially when such acts involve excesses by the military establishment upon the helpless, whether military or civilian, the resulting outrage by the media can reach firestorm proportions. Military reprisals against civilians, even in times of war and against enemy civilians, are repugnant to our value system and something that will not be tolerated by the United States citizenry.

Occurring as it did sometime during the first week of July 1947, the Roswell Incident happened at the time of the first wave of flying saucer sightings around the country that summer. The latest "sighting of the day" commanded front-page attention in most newspapers, as was the case with events in Roswell. As an anxious and excited nation—and world—awaited more news of the discovery, things were about to change. Moving quickly to kill the story, our government used a combination of appeals to patriotism, claims of "national security," bribery, threats of long prison sentences, and outright thuggery in the form of death threats to contain the story. As a result, the Roswell Incident turned into a two-day story and was quickly out of the public consciousness.

Those in the military who were involved in the retrieval of the wreckage, the bodies, and what was left of the crashed UFO itself were the easiest to deal with. Roswell Army Air Field was a SAC base, so everyone who worked there, military and civilian, was already familiar with the base policy of not talking about things that went on at the base, even to family members—ever. To drive home this point, the enlisted men involved in the cleanup at the various sites were detained in groups and "debriefed" (sworn to secrecy under the guise of national security). Long prison terms were promised in case anyone was thinking of talking, and we have also heard that bribes of $10,000 or more were used to assure the silence of those who saw the bodies. The *officers* involved, especially career officers, were less of a problem. In order to advance a career in the military, one does not defy orders or breach security. One key officer who was heavily involved in the recovery operations even promised President Truman (via Truman's aide) that he would keep the secret forever. He did, until he was on his deathbed many years later. Controlling civilians, however, was a different matter.

Except in time of war or under conditions when martial law has been declared, under the Constitution, the U.S. military has no direct authority over U.S. civilians. The military could keep its own house (the men of 509th Bomb Group stationed in Roswell and up the chain of command) quiet, but how to keep the civilians from exercising their God-given, Bill of Rights–guaranteed freedom of speech? And there were a lot of civilians involved in the Roswell Incident all along the way: from the initial discovery of pieces of wreckage by civilians near Corona, to the discovery of the craft itself closer to Roswell, to the recovery operations at the Roswell base itself, and finally to the shipment of the wreckage and bodies to Wright Field in Dayton, Ohio.

Aside from the Corona ranchers whose homes were indiscriminately ransacked[1] in our military's mad search for "souvenirs" from the crash, and the rancher Mack Brazel who started it all and who was dealt with directly by the military, other civilians involved in the 1947 Roswell events were dealt with through civilian authority figures. The highest-ranking of these was Dennis Chavez, a U.S. senator from New Mexico. He was "enlisted" by the Army Air Forces to intimidate Roswell radio station KGFL, whose ownership had secured an exclusive, recorded interview of Mack Brazel, during which Brazel told of finding strange wreckage and the bodies of "little people." Walt Whitmore Sr., the station's majority owner, was threatened by Senator Chaves with the loss of the station's broadcasting license if it went ahead with its plans to air the Brazel interview (KGFL had planned to "scoop" the other Roswell media outlets with the interview). KGFL minority owner Jud Roberts was also similarly threatened, and for good measure, the head of the Federal Communications Commission in Washington, D.C., "T.J." Slowie, also threatened Whitmore and Roberts with the same message.

It worked better than was hoped for, as Whitmore caved in completely by not only turning over the taped Brazel interview, but also becoming a willing accomplice in the military's campaign to silence civilians. One of his broadcasters, Frank Joyce, had also been called by "a military person in Washington" and had been "read the riot act" to shut up about the crash.

Walter Whitmore Sr. (Photo courtesy of Mrs. Walter Whitmore Jr.)

Joyce had been the first media person to interview Brazel and knew Brazel's original story, including the part about him finding little bodies. Incensed about being told what to do by someone in the military, Joyce let him know where to go. The angry voice in Washington shot back, "I'll show what I can do!" and hung up. A day or two later, Joyce's boss, Walt Whitmore Sr., told Joyce to get into his car, and that they should go for a ride. Joyce did so, and then noticed a strange-looking man in a strange-looking uniform sitting in the back seat. The man did not speak. Whitmore drove north out of Roswell for more than an hour to a remote shack off Corona Road. He was told by Whitmore to get out of the car and go into the shack. This Joyce did, still not knowing what was going on. Joyce stood alone in the shack for a few minutes, wondering what was taking place, when in walks none other than Mack Brazel himself. "You're not going to say anything about what I told you the other day, are you?" Brazel asked Joyce. "Not if you don't want me to," responded Joyce. "Good. You know our lives will never be the same." With that, Brazel walked out, and Joyce never laid eyes on him again. Joyce then returned to Whitmore's car for the ride back to Roswell, and the stranger in the back seat was gone. Apparently the military was not sufficiently convinced of Joyce's pledge not to say anything, and he was shortly thereafter gathered up and physically removed to a Texas hospital for a year or so under circumstances that were never clear to him. A Roswell native, Joyce did not return there upon his release.[2]

We know of at least one instance when the military authorities took matters into their own hands without the use of surrogates by employing thuggish tactics directly to civilian eyewitnesses to scare them into silence. Perhaps they thought they could get away with it because one of the witnesses involved was a mere child. Perhaps they hadn't gotten their game plan together yet. Frankie Dwyer was a twelve-year-old schoolgirl in July of 1947, and her encounter with the bullying and threatening military officer haunts her to this day. Frankie's father was a crew chief with the Roswell Fire Department. At the time of the incident, Frankie had just been to the doctor after having her tonsils removed the week before and had stopped at the fire station to wait for her father to take her home. While she was waiting, a highway patrol officer by the name of Robert Scroggins came into the station with a mischievous look on his face. "Hey, guys! Want to see something?" Scroggins reached into his pocket and took out something that he wadded up in his hand. "Watch this," he said. He then held his hand out about a foot over a nearby table and opened it. Something silvery fell from his hand to the surface of the table without a sound. Then, as if by magic, it spread out like quicksilver into a small, thin, irregular sheet of *something* in one or two seconds. Everyone who was there that day, including Frankie Dwyer, had a chance at it. It could not be cut with scissors, scratched, burned, or permanently creased. No one there could figure out what it was. Scroggins said that he had gotten the piece from "someone in Corona."[3] A day or two later, Frankie Dwyer was at home with her mother tending to chores when there was a hard knock on the door. Opening the door revealed a tall man with wide shoulders and a dark complexion dressed in an MP uniform. He was looking for Frankie Dwyer. After Mrs. Dwyer introduced her daughter to the ominous MP, two other MPs escorted Mrs. Dwyer into another room. With a thick New York accent, the first MP started to question Frankie about the incident at the fire station and wanted to know what she had actually seen. Satisfied that the twelve-year-old had seen plenty, he took out his billy club and started pounding it into his open palm as he tried to pound home his reason for the visit. "You did not see anything. You got that? If you say anything, not only will you be killed, but the rest of your family will be killed too. There's a big desert out there. No one will ever find you."[4] With that, a shaken Mrs. Dwyer was returned to her terrified daughter, and the men left. Meanwhile, back at the Roswell Fire Station, Philbin's boss at the RAAF, Colonel Walter C. Lucas, was applying the pressure. Earlier in the day, the firemen had been paid a visit by Roswell City Manager C. M. Woodbury, a former Army major who was very tight with Colonel Blanchard. Woodbury had heard about the department's trip to the crash site north of town and had come to the station on instructions from Blanchard to tell the men to say nothing about it. But he hadn't known about the visit by the highway patrol officer and his memory-metal demonstration for everybody there. Things were getting out of hand. Hearing this from Blanchard, Colonel Lucas decided to take matters into his own hands and go to the fire station himself, because his chief security officer, Major Robert Darden, and Lieutenant Philbin were already otherwise occupied. According to a fireman who was there, Colonel Lucas's message was a stern warning, simply and forcefully delivered, "You are to say nothing—*ever*—to anyone about this!" That was all. No questions were solicited or asked.[5]

In researching Frankie Dwyer Rowe's story throughout the years, we followed the evidence trail to conclude that the MP who most likely confronted Frankie that day was a former Brooklyn, New York, policeman by the name of Arthur Philbin, who was a security officer with the 390th Air Service Squadron that was part of the 509th Bomb Group in 1947. Lieutenant Philbin, in addition to being tall and dark with wide shoulders, ran the guardhouse on the base and had a reputation for being an all-around tough guy. He also was the base liaison with the Roswell Police Department and Sheriff's Office on matters that affected both the base and the town. Philbin died many years ago, so it was impossible to turn the tables and interrogate him about his Roswell days. Therefore, we never published anything about Philbin or anything that mentioned his name. We also never mentioned his name to Frankie Dwyer Rowe. In 2005, we got an idea: Why not try a lineup similar to what the police do when they parade a number of people in a group that includes the suspect before the witness? The hope is that the witness will be able to pick out the suspect from the rest of the group. Sometimes the witness can, and sometimes not. No harm in trying. We knew that Lieutenant Philbin's picture was in the 1947 RAAF yearbook. He is shown on a page along with the pictures of sixteen other officers—enough for a lineup by any standard. We then made a photocopy of the pertinent page from the yearbook and mailed it to Frankie Dwyer Rowe with the simple question: "Do you recognize anyone on the enclosed page as the person who came to your house and threatened you back in 1947?" A few weeks later, we received an envelope in the mail bearing Frankie Rowe's return address. Inside the envelope was the folded, photocopied page that we had sent to her. There was no accompanying letter, just the page with the pictures of seventeen officers of the 390th ASS on it. There was simply a single circle drawn around one of the pictures—the picture of Lieutenant Arthur Philbin.

In the summer of 1947, Sue Farnsworth saw such a frightened look on her father's face as she had never seen in her young life. After all, Arthur Farnsworth was a successful and prominent businessman who owned and managed Roswell's original Ford dealership, Roswell Auto Co., located at the intersection of West Second Street and Richardson Avenue in downtown Roswell. He was a pillar of the community. Besides their home at 612 North Richardson in town, the Farnsworths also owned a working ranch northwest of Roswell that the family also frequented as a "getaway." Arthur, especially, was known to visit the ranch several times a week. Seven-year-old Sue Farnsworth had two older sisters, both of whom had polio, and it was Sue who became best buddies with her father by helping out with the ranch chores and sharing important father/daughter matters with each other when the need arose. "We were used to finding 'funny things' from White Sands on our ranch all the time, but this was clearly something different judging by the military reaction."

Unable to contain her concern for her father any longer, Sue finally asked him what was wrong one day when they were out at the ranch. Without saying a word, he motioned to her to follow him. He then mounted a horse, as did she, and they rode out to a remote and secluded part of the ranch, whereupon they dismounted. "Whenever we wanted to discuss something without anyone else hearing, we would go to this spot." Like everyone else, Sue had heard the stories and rumors about the UFO crash that were rife at the time, but she was not prepared for what her father was about to confide to her. Like so many times before, Sue and her father sat down on their favorite rock and stared out over the peaceful, scrub-desert landscape. "Your father was threatened by the military a few days ago," Arthur Farnsworth told his daughter. Looking around carefully even though they were out in the middle of nowhere, Arthur Farnsworth continued in a low voice, "What I tell you now you must never tell anyone. A *flying saucer* crashed on another ranch near here, and the military told us that if we ever said anything to anyone about it, they would kill all of us. I went out there with some other ranchers soon after word of the crash got around, and we saw some things we weren't supposed to see." Arthur Farnsworth did not go into any detail with his young daughter. "Remember, not a word—to anyone!" That was all he would ever reveal to her. The two then saddled up and rode back to the ranch house, never to speak of the matter again.[6]

We can speculate now what probably happened to Arthur Farnsworth to cause him to fear for the lives of himself and his family. He and the other ranchers he was with must have gotten a fairly good look at not only the physical crash wreckage, but also the alien bodies. Terminal threats were reserved for civilians who had witnessed the latter. Most likely, Farnsworth had visited one of the body sites (the "Dee Proctor Site" on the Foster ranch in Lincoln County or the "impact site" in Chaves County) shortly after the crash on one of the days that he was out at his ranch. News of the crash traveled like a flash among the ranchers even in those days, and ranchers with their excited children always got to airplane crash sites before the authorities. And it was no different with the crash of a UFO. Upon hearing news of the crash, Arthur Farnsworth, with his neighboring ranchers, headed for the crash site to see for themselves. This would have been sometime between July 4 (the day following the crash) and July 7 (the day the military started securing the crash sites with armed MPs).

When we interviewed him, George Cisneros still lived on the same ranch he lived on in 1947 near the town of Arabela, New Mexico, which is about twenty miles from the Foster ranch where Mack Brazel was the foreman. According to Cisneros, who did not go to any of the UFO crash sites himself, a number of his neighbors *did*. When we asked him if his neighbors reported seeing anything, his response was, "Hell, yes! When they came back, everyone was excited and buzzing about a crashed spaceship and 'little space people.'"[7]

If Arthur Farnsworth was among them or with a different group of ranch-ers who visited one of the two crash sites with bodies, he no doubt saw the same things. Perhaps the military arrived at the crash site soon thereafter and issued a stern warning to everyone, including Farnsworth. Or, perhaps a few days or weeks after that, he was paid a visit by another military officer, such as the "brutal" Hunter Penn (see Chapter 6 and the postscript at the end of this chapter) whose job it was to conduct a "sweep" of the ranches near the crash sites to enforce the silence of those ranchers who knew too much about the crash, and especially about the bodies. As we have seen, the threat of "terminal action" was authorized to secure compliance. In the words of Sue Farnsworth, who spoke to us sixty-two years after the fact, "I never in my life saw my father so scared, before or since."[8]

In the summer of 1947, Richard Loveridge worked as a mechanical engi-neer for the Boeing Aircraft Company. In that capacity, he was part of Boe-ing's aircraft crash investigation team, working the states of Texas and New Mexico. For this reason, he lived part of the time in San Antonio, Texas, and part of the time in Roswell, New Mexico.[9] In those days, Boeing was *the* major supplier of heavy bombers to our country's military arsenal. The company's B-17 Flying Fortress contributed mightily to the defeat of Germany during World War II, while its B-29 Superfortress did the same in the Pacific The-ater of Operations in the defeat of Japan. It was two B-29s named *Enola Gay* and *BocksCar* that finally ended the war with the atomic bombing of Hiro-shima and Nagasaki. Boeing also supplied a number of non-bomber military aircraft used in various support roles. For instance, its C-97 Stratofreighter, introduced in 1947 as a cargo plane, soon became the KC-97 Stratotanker for the purpose of refueling long-range bombers in flight. Today's heavy bomber fleet includes two Boeing models, the B-1B Lancer (ironically flown by pilots of the 509th Bomb Wing—the direct descendant of Roswell's 509th Bomb Group!—out of Whiteman AFB, Knob Noster, Missouri), and the venera-ble, nearly seventy-year-old B-52 Stratofortress. Given the large number of Boeing models in operation at any given time in our nation's history during and since World War II, it stands to reason that some of them would crash, and it is therefore entirely reasonable that Boeing, the designer and manu-facturer, would want to have its own investigative team on the scene when one of them did. And in the summer of 1947, Fort Worth Army Air Field (FWAAF) was the headquarters of the Mighty Eighth Air Force of World War II fame, which helped bomb Germany into submission, while Roswell Army Air Field was home to the 509th Bomb Group, the outfit that dropped the two A-bombs on Japan to end the war. Its aircraft of choice was the Boeing B-29 Superfortress heavy bomber.

Douglas Loveridge is an aeronautical engineer who has worked on classi-fied government programs for over thirty years for the Northrop Grumman Aircraft Company in Oklahoma. He has also worked on new experimental air vehicles and new avionics systems for ten years at Edwards AFB in Cali-fornia. He is also Richard Loveridge's son. Knowing what his father did for a living and that he had lived in Roswell for a period of time in the 1940s and 1950s, when *The Roswell Incident* book was published in 1980, Douglas put two and two together and asked his father the "Daddy, what did you do in the war?" question. "I don't know anything. Don't ask me about it. They can hurt you!" was his father's curt response. In the ensuing years, whenever Douglas would periodically ask his father about Roswell, he would notice a big change in his father's demeanor, that he would become extremely agitated. And the answer was always the same: "Don't ask me. They can hurt you!" At some point, however, his father started to soften his position in response to his son's persistence. "Yes. I heard the crash stories at the time, but, thinking that one of our bombers had crashed, I drove out there [to the UFO crash site] but was turned away [by the military]. I didn't get to see anything. So, I still don't know anything." Then, in the last year of his life, in 1993, Richard Loveridge finally relented and admitted that he saw everything. When asked about it yet again, he didn't get agitated this time. He said matter-of-factly that he had been to the crash site north of Roswell in his official capacity as a crash investigator and that he saw the wreckage, as well as three small, deceased "entities" and one that was still alive. He said that the "entities" were "child-size" and "grayish" in color. Nothing more. Finally, becoming emotional now (according to his wife, Richard Loveridge had a nervous breakdown at the time of the incident), he repeated the all too familiar warning, "Don't ask me any more about this. They can hurt you!"[10]

Chaves County Sheriff George Wilcox was the local authority figure utilized by the Army on the ground in Roswell to help contain the story. A picture of him on the telephone and looking like a deer caught in the headlights was prominently featured on the front page of the July 9, 1947, edition of the *Roswell Daily Record.* But Wilcox was doing more than just answering telephone calls. He refused to give out any details to inquiries regarding what was going on, because he was "helping out the fellows from the base."[11] He also completely rolled over in the face of the military full-court press by allowing himself to be used as the enforcer to intimidate local Roswellians into keeping their mouths shut about what they witnessed.

Chavez County Sheriff George Wilcox. (Photo courtesy of Mrs. Phyllis Wilcox McGuire.)

It was his task to deliver the threat of the ultimate sanction to those who saw or knew about the bodies recovered from the crash. Glenn Dennis, the Roswell mortician, allegedly knew about the bodies from a nurse friend who was involved in the autopsy of one of them at the base. Dennis had been threatened with death at the base hospital by an officer, and, similar to Frank Joyce, Dennis became incensed at his treatment by the officer and told him where he might go. The next day, Dennis's father received a visit from Sheriff Wilcox and a deputy to tell him that his son "was in trouble at the base."[12] No doubt the veiled death threat of the previous day was part of the message delivered by Wilcox.

Ruben and Pete Anaya in 1947 were Montoyistas, young supporters of then lieutenant governor and future New Mexico senator Joseph Montoya.

After the incident that shook him up so badly, the two Anaya brothers and their families were paid a visit at their homes by Sheriff Wilcox. According to Pete Anaya and his wife, Mary, who were interviewed again in September 2002 about this incident, Wilcox delivered the ultimate sanction to them if they talked about what they knew: "If you say anything, you will be killed. And your entire family will be killed as well."[13] It is not known to how many others Sheriff Wilcox delivered the message on behalf of the Army, but what is known is that he never ran for sheriff again. According to family and friends, the

Wilcox's granddaughter Barbara Tulk Duggar. (Photo courtesy of Tom Carey.)

Roswell events "destroyed him." Now we know why. When asked about all this just a few years ago, a former deputy of Wilcox's responded, "I don't want to get shot." After the story had been contained and things had died down a bit, the Army paid a visit to Sheriff Wilcox and his wife, Inez. To praise or reward him for a nasty job well done, you may ask? Think again. The message delivered to the startled couple was that unless they kept quiet about everything, not only would they too be killed, but their children would also be killed. Sheriff Wilcox died in 1961. Asked by her granddaughter years later whether she believed the threats or not, Inez Wilcox looked at her with a straight face and clear eyes, and said, "What do you think?"[14]

POSTSCRIPT

In Chapter 6 we described for you the "dirty work" activities of the Army Air Forces officer Hunter G. Penn. You will recall that Penn was brought to Roswell in the weeks and months following the UFO crash retrieval to enforce the continued silence of the ranchers northwest of Roswell who might have seen things, some of whom, it was felt, might have retrieved "artifacts" from the crash for themselves. Collecting souvenirs from airplane crashes had been standard procedure throughout the years for the ranchers, especially for their children,

Hunter Penn, Air Cadet in Officer Candidate School in the early 1940s. (Photo courtesy of RedSun.)

when crashes were encountered by them, and we believe that artifacts from the 1947 UFO crash/retrieval are still out there.

While she was growing up, Hunter Penn's daughter, Michelle, said she was fearful of her alcoholic father ("He was a 'brutal' person," she said), and she didn't need to be told never to say anything about what he had recounted to her about his involvement at Roswell in 1947. To this day, she is not sure why he told her, but it may have been his way of parenting, to use a fear factor to instill a desired result in the behavioral control over a child. She said that she was made to address him as "Sir" as a child. According to Michelle, her father was also "brutal" with her alcoholic mother, Elinor. The bottom line here is that Michelle Penn believes that her father was entrusted with the role of being the "Bad Cop" when the situation called for it, because he was known as being brutally tough with people—just what was needed at Roswell. In more dramatic terms, she said, "He tried to 'heart-attack' people!" Michelle said that her father would sometimes brandish a military pickaxe (similar to an ice pick). She thinks that he was obsessed with picks and believes he may have used one at Roswell.[15]

19

THE FLYING SAUCER TAKES
A TRAIN FROM ROSWELL

I f one were to examine a plat of the former RAAF in 1947, you would note that one entrance to the base proper which provided the utmost security from observation was the east gate. Since we began our investigation in 1989, we have had numerous civilian witnesses describe the convoy transporting the remains of the crashed ship detour from downtown Roswell on Main Street, onto Second Street turning to the east side of town, and from there taking a right at Alexander and proceeding south, which would have them arrive just east of the perimeter on that side of the base. The cargo could then enter through the east gate, cross the open area B-29 runway, and go directly to Hangar P-3, precisely where everything related to the crash was being delivered. As a matter of fact, there was an entire roll of hangars and aircraft service buildings along the east side of the RAAF which served as a total wall to provide total secrecy. And no one on the base would have been able to see any such clandestine arrival throughout the entire recovery operation. It was the perfect plan. As previously described it was from this building that everything remained under heavy guard before being crated up and flown out to other destinations—most notably Wright Field. Additionally, we have secured the testimony of numerous personnel normally assigned to P-3 who described being ordered not to report to the building for days during the "flying saucer" event. Everyone had suspicions as to the nature of the quarantine but more important was the fact that lethal force was implemented for anyone who would violate that shutdown.

Now, running parallel to the east fence line just another hundred feet outside ran the old Southwest Pacific rail lines. From there the tracks merged with rail systems that spidered throughout the United States, in this case from New Mexico, through Texas, then up through Kansas and as far east as St. Louis. From there a separate line would have transported any special cargo into Ohio.

Unlike many military bases that were constructed over operational train tracks, which enabled them to regularly use the freight lines for shipping purposes, the RAAF was a top-security base and could not allow for an open railway passing directly through the facility. For that reason, the base was situated with the tracks just outside the secured area and, in this case, just outside the east gate.

To facilitate the base's use of the train, a freight yard was set up to accommodate special loading and unloading operations. As the first strategic air command base in the world, the RAAF conducted such exercises within standard security measures. MPs would often accompany the transporting of sensitive shipments until reaching their final destination. So it should come as no surprise that numerous witnesses to the events surrounding the incident of 1947 have mentioned the use of the railroad to ship some of the physical evidence out of town.

As it often has been demonstrated how many roads from Roswell led to the state of Ohio, the following account, if not for numerous stories involving the Roswell freight yard, would remain just another anecdote. We'll let the reader decide.

Ralph A. Multer received a Purple Heart medal after serving in the Navy during WWII. After leaving the service, he became a truck driver at the Timken Company of Canton, Ohio. Timken was heavily contracted during the war, providing ball bearings and steel tubing for military applications. Timken also was one of the largest wartime producers of large gun barrels. The company possessed major security clearances with the Pentagon, which remained in effect after the war. It also had a metallurgical furnace, a blast furnace that was one of the hottest in the world—more than 2,000 degrees Fahrenheit.[1]

On a number of occasions, we have heard how some of the GIs at the 1947 RAAF sneaked some small samples of the wreckage from the crash to a garage just outside the base. There, they had an auto mechanic try to deform a piece with a welding arc. Just as with every other described attempt, even an acetylene torch couldn't affect the metallurgical characteristics of the strange material.[2] If such accounts are true, and based on all the testimony that paint visions of nothing but failed attempts at cracking the fracture point of "the stuff," it is reasonable to assume that they would have tried extreme heat. No doubt, Timken would have provided a facility to conduct such testing. Unlike the steel-making center of Pittsburgh, Timken made its own steel to insure supply and quality control. And just as important, it was heavily contracted by the U.S. government, and it was just down the road, so to speak, from Wright Field. Something had to break the material's molecular code, and Timken could blast it with extreme heat like no other.

Multer, who had a security clearance with Timken back in 1947, apologized to his wife, Violet. Instead of meeting her for lunch upon completing his shift, as he routinely did, he was sent out to the railroad yard with his truck. A special cargo had arrived from New Mexico and it needed to be immediately hauled over to the Timken plant. Something covered by a canvas tarpaulin was loaded into the back of his truck as men in suits, purportedly with the FBI, oversaw the procedure. All of this only added to the growing mystery of what the secret cargo was and why it was headed to Timken. Whatever it was, it was to be tested for strength and durability in a super-hot Timken furnace.

"They talked to a person later who was there that night and they said they couldn't cut it. They couldn't even heat it," said Sundi Multer-Lingle, Multer's daughter. She added, "The piece of metal—well, I don't know if you can call it metal, the object was absolutely impenetrable."[3]

According to Multer, one of the FBI types made it very clear that not a word of the clandestine affair was to be spoken. Still, Multer would confide in his wife about it and, starting in the 1960s, would describe to his daughter the day he "hauled material from the crashed spaceship." It became a well-kept secret within their family until Ralph passed away around 1990. It was then that Violet and Sundi spoke out publicly about the lightweight metallic, silver-gray material that Timken couldn't melt, dent, or even scratch. His wife summarized, "The experience left a lasting impression on Ralph. It was always on his mind."[4]

What makes the previous information plausible is the story we are about to relate. Keep in mind that some of the crash debris allegedly arrived in Ohio via rail. We now return to Roswell just outside the east gate—just before the train leaves the station.

As the background picture was painted earlier, the freight yard in Roswell in 1947 was just outside the east gate. For that reason it fell under the jurisdiction of the federal government under the authority of the civil service. The supervisor of that specific loading area was the late Charles Austin Wood, who is listed in the 1947 Roswell City Directory as a federal employee working for the civil service.[5]

Within days of all the talk about the crash of the flying saucer, Wood checked his morning freight assignments, which included a shipping car for the RAAF. No destination was given, and Wood would need to be present and then sign off that the cargo was loaded, secured, and en route to its next stopover. He noted the time for departure and arrived earlier to ensure an available empty rail car.

At the predetermined time, a jeep of MPs raced across the tarmac proceeding toward the east gate. About 10 feet from the heavy wire-mesh fence, two of the guards leaped from the rear seat, and as one unfastened the padlock securing the gate, the other started to slide it along a metal channel, clearing a space for vehicles to drive through to the outside. Shortly thereafter, a truck followed, exited through the same opening, and arrived at the waiting freight car. The armed MPs next flanked the perimeter as the truck was guided and it backed toward the open car. Wood continued to watch as soldiers began to take small wooden boxes from the rear cargo area of the truck and slide them into the waiting sidecar. An officer also observed and instructed the men's activity. The civil servant noted that the boxes were open at the top and appeared to contain pieces of jagged metal. Out of curiosity, Wood contemplated some plane mishap, but none had been reported in the local news. The base was typically very up-front with its civilian neighbors regarding such incidents. But why all the security for mere wreckage? The pieces were all similar in appearance from Wood's vantage point . . . all looked like metal that had suffered some form of fracture.[6]

One of the actual MPs who took part in this situation was PFC Frank Vega, with the 322nd Troop Carrier Wing who, with five other recruits, was flown down to Roswell from Lowry Field in Denver, Colorado. According to Vega, they were issued "grease guns" and assigned to guard Hangar P-3 with the orders "Don't ask any questions" and "Shoot to kill" any unauthorized individuals attempting to enter the hangar. Vega confirmed that the entire base was on "lockdown" and that he and his companions spent a full week, which included the loading of wreckage from the hangar being "loaded onto freight cars at the railroad spur."[7]

The workers were getting down to the last crate, and some of the enlisted men hopped back into the truck. Wood's curiosity continued and then something unexpected happened to reward his patience. As one of the last boxes was lifted into the car, a single remnant fell off the side of the container and dropped partially obscured under the boxcar. The soldier didn't seem to notice, as he crawled into the back of the truck just as the rear tailgate was slammed

Frank Vega in an early 2000s photo at the front door of his home. (Photo courtesy of Tom Carey.)

shut. Within moments the vehicle was heading back toward the east gate. The project complete, the officer waved Wood over and scribbled his signature on the work order sheet without uttering a word. And when the entire military team had cleared the area and driven out of view back onto the base, Wood casually walked over to the now-closed freight car and stepped on the orphaned piece of metal down in the dry, dusty ground with his shoe. So far, so good. "Take 'er away!" shouted Wood to the engineer as he built up steam.

Railroad boxcar near RAAF East Gate in a 1990s photo. Now gone. (Photo courtesy of Tom Carey.)

That night as Roswell slept, Wood parked his car some distance from the east gate and silently slipped back to the train yard. Unlike the base's main front gate, there were no lights or sentries to worry about. The same remoteness and lack of human observational presence now served him well in his own recovery mission. After gaining his bearings and then feeling around the ground alongside the metal rail tracks in total darkness, there it was: just beneath a thin layer of dry dirt. The orphan piece of metal now had a father by the name of Wood.[8]

Serviceman Frank Vega would tell us that after they completed the guard duty at the hangar, they "smelled like sick skunks."[9] Rather than return to Lowry, they were first sent to Amarillo Army Air Field in Texas, which had just been shut down on September 15, 1946, and whose buildings were converted to peacetime uses or demolished after WWII. For the next six months they were detained there and not permitted to leave or have any outside communication during that entire "layover."[10] The entire Roswell affair needed to disappear, including the people who could talk.

Days would pass, then weeks, and after a few months it seemed like Wood had managed to accomplish what so many others had failed to do. There were no strange visitors, no unexpected intruders, no one shadowing him about the community. Charles Wood had in his possession something that made all the rumors true . . . something that defied the news reports that said all the excitement was simply over a weather balloon. All those witnesses like Frank Vega, who knew otherwise, were forced to remain silent. Should Wood have become an obscure member of their ranks? The difference was obvious: the authorities didn't know about or suspect him. And what hornet's nest does one disturb when you are in possession of something highly classified? As a government employee, Wood knew the security regulations and the penalties. And what if they were to accuse him of stealing it? What to do with it? Tell the world or keep it secret? Readers will have to debate his decision.

Life in Roswell went on. Nevertheless, as the years passed, people still whispered among themselves. An air of lethargy remained within a community in which the high hopes for the post-war period were now tainted with suspicion and doubt. Wood wisely avoided the need to live down the stigma of making claims that couldn't be proven. Still, he kept his evidence concealed to himself. Not even his family knew what deep secrets he kept buried, as if still hidden out at the old freight yard.

The Woods' son James anxiously awaited the celebration of his sixth birthday in 1952. What could possibly be in that small box simply wrapped with no ribbon or bow? Not James, for that matter, not anyone, would ever have guessed, even after he opened the present and looked at it for what seemed like minutes waiting for his father to tell him what it was. Charles finally asked, "Do you remember when I told you about all that commotion a few years ago when they thought a flying saucer crashed outside of Roswell? That's a piece of the weather balloon."[11] But James quickly realized that his dad was joking, because the more he folded it and rolled it into a ball as Charles instructed, it always unraveled back to its original shape. "Now watch this," Charles exclaimed. He then flicked his cigarette lighter and held the piece over the flame until it should have been red-hot. Still, there was no effect as the boy's eyes grew larger and larger with each new demonstration. "Where did you get this?" For the first time James listened intently as his father recalled every detail back in 1947 at the freight yard and the lost piece of wreckage. "I thought they would have missed it by now," confessed Charles. "Now it's yours. Happy birthday, son."

Soon James realized he possessed something special . . . something none of his playmates had. The very possibility that he owned an actual piece of a flying saucer both frightened and fascinated James. And as the days passed, he felt all the more compelled to share it with his friends. The dilemma was how to display the piece of metal in all of its uniqueness? How to fully highlight its advanced characteristics? It wasn't long before James arrived at a most novel solution.[12]

Like so many other boys his age, James was developing a childhood curiosity with the art of magic. In fact, he began to test his own skills in an attempt to perform his own tricks. His father said that he and his neighborhood friends could use an old wooden shed as their clubhouse, which could also serve as James's private magician's parlor. And wouldn't you know it he already had the perfect finale for his routine.

Soon, little James Wood became the talk of his small, local community. The young magician was drawing even adults to his act—as many as could fit into the old wooden shed. And each and every time, James would end his performance with the same climax: he would take the tissue-sized piece of metal from his folding table, display it to his audience, crunch it into a small ball, say some magic words, and finally toss it in the air just in front of where he stood. All watched in amazement as it unraveled and softly floated to the ground . . . each and every time. Applause, applause, as James quickly snatched up his star magic prop to secure it away for the next show.[13]

James Wood never let on how he happened to acquire the mysterious piece of metal that seemed to have a mind of its own. He never explained where it was from. After all, magicians never reveal their secrets. Unfortunately for James, in fact, for all of us, someone else was watching who decided the show needed to end. And as one could have predicted, late one night, while all were asleep, someone broke into the locked clubhouse and whisked away the piece of metal. Only the Woods family knew of the true provenance of what locally was nothing more than an unusual stage trick. Nevertheless, someone recognized the characteristics as something beyond magic . . . something potentially unearthly. And it was worth the risk of stealing it even from a child. A number of Roswell police officers went so far as to suggest that the stolen artifact was most likely in Washington, D.C., by the time they were notified.[14]

Throughout the course of our investigation, we have learned of many attempts to understand and analyze the "magical" properties of the wreckage from Roswell . . . some of the testing involved tremendous heat. As we have demonstrated, some of the material was shipped from Roswell via freight train with a possible destination of a metals plant in Ohio that at the time had one of the hottest heat furnaces in the world. Aside from that, why Timken? What else did the company provide unlike any other of its kind? We offer what we believe is a most significant piece of deductive evidence:

At the turn of the 20th century, Stephen Timoshenko was a Russian genius in mechanical engineering and materials composition. More over, he was a super-genius for whom metals used in engineering were named. Timoshenko migrated to the U.S. and soon thereafter joined Westinghouse Electric. From there he taught engineering mechanics at the University of Michigan and would later retire from Stanford. To this day he is considered the father of modern engineering mechanics.[15]

While at Michigan, one particular student not only stood out as a prodigy but was years ahead of his fellow classmates. Oscar Horger would write his dissertation, *Effect of Surface Rolling on the Fatigue Strength of Steel* and receive his PhD under Timoshenko in 1935. He would become so renowned on the subject that he wrote several manuals for the American Society of Mechanical Engineers (ASME) on metals, engineering, and design and later would become the ASME president.

It was immediately after acquiring his doctorate that he put his outstanding knowledge to work with a company in Ohio. There, in 1937, he established an industry-leading laboratory, which he directed and was on the cutting edge on such topics as fatigue-testing of materials, strength of metals, and measurements of surfaces. This last category included a 5000X magnifier and a new device called a profilograph used in high-tech surface measuring at the molecular level.

In 1940 Horger would move to Stark, Ohio, and become the chief engineer of the same company's railway division conducting fatigue investigations of railroad axles and crank pins. Before 1947, Dr. Oscar Horger was known around the world as the "authority on testing metals." And the company with whom he earned that title and where he was working at the time of the Roswell Incident—Timken.

20

"IT WASN'T OURS!"

When Corona sheep rancher Mack Brazel heard a strange, muffled explosion among the thunderclaps during a severe lightning storm late one evening in early July 1947, he had no idea what it could be. ("The lightning seemed to be attracted to a single location on the ranch," he would later tell his son Bill.) Some of Brazel's neighbors also told of hearing the explosion. The following day, as he gazed upon one of his pastures now covered by pieces of silvery wreckage, he still had no idea what it was. Without electricity or a radio and with only monthly newspapers, he had not heard about the flying saucers that had made their historical appearance only two weeks previously. It wasn't until Brazel had paid a visit to his close neighbors the following day, and to a Corona bar a few evenings later, that he heard about them—as well as a possible reward for finding one. Early the following morning, he would find something else on his ranch, something that convinced him to drive to Roswell to report his finds, thus securing his place in history and initiating one of the greatest mysteries of the 20th century.

Leaving the something else for another day, what was it exactly that was deposited in small pieces concentrated so densely in a fan-shaped area of up to 1½ million square feet in Brazel's pasture that summer of 1947? From various sources, both military and civilian, who actually saw and handled pieces of the wreckage, we have determined the most interesting consisted of the following:

1. A large amount of small- to palm-sized pieces of smooth, very thin, very light but extremely strong "metal" the color of aluminum, which

could not be cut, scratched, bent, or burned. Some of the larger pieces displayed a slight curvature to the surface.

2. An apparently large quantity of palm-sized and larger pieces of a very thin and very light "metal" or "cloth" with "fluid" properties. This type of wreckage also could not be cut, scratched, or burned, but it could be temporarily deformed. As startled witnesses tell us, "I wadded up a piece of it in my hand, and it felt as though there was nothing there. Then, I placed it on a smooth surface, and it unfurled itself and flowed over the flat surface like liquid mercury back to its original shape without so much as a crease." It is this so-called memory metal to which we refer today as the Holy Grail of our Roswell investigation, because locating a piece of it would provide irrefutable proof, we believe, of the extraterrestrial nature of the Roswell Incident.

3. A quantity of thread-like, monofilament "wires" that could not be cut, scratched, burned, or permanently deformed. These could be coiled, and when a light from a flashlight was shown in one end, it could be seen coming out the other end. These, it has been suggested, adumbrate today's fiber optics technology.

4. A small, black, seamless box that could not be opened.

5. A light, seamless, dull aluminum flange or strut that could not be cut, scratched, or burned.

6. A number of thin "I-beams" about 18 to 30 inches long by ¼ inch wide by ⅜ inch thick containing "writing" in the form of unintelligible symbols along their lengths; these "beams" were as light as balsa wood, but they could not be cut, scratched, or burned; they could be flexed slightly but not broken.

7. A number of thin strips of thin, aluminum-like "metal" about 3 to 4 feet long by 3 to 4 inches wide containing a hieroglyphic-like "writing" on them.

It should be noted that debris similar to what has just been described was also found at the impact site in proximity to where the remainder of the intact craft allegedly crashed—but not in the large quantities found in

Brazel's sheep pasture on the Foster ranch. So, what can we make out of such items as those just described? Were they truly as exotic in nature and origin as they appear, or just misidentified everyday items (tinfoil, sticks, and rubber), as the U.S. Air Force would have us believe? To answer this question, let's let the people who actually handled the materials tell it like it was.

We know that Mack Brazel told the *Roswell Daily Record* on July 8, 1947, that what he found was certainly "no weather observation balloon." Later, he confided to a few family members that it was "the strangest stuff he had ever seen," and, according to his son Bill, the Army admitted to his father that they had definitely established that "it wasn't anything made by us."

The RAAF base intelligence officer, Major Marcel, thought from the very first, and until he died in 1986, that the wreckage was from an extra-terrestrial spacecraft. His son, Jesse Jr., confirmed to us that his father was already talking about the wreckage in terms of it being from a flying saucer during his early morning visit home with some samples prior to reporting back to the base from his trip to the Foster ranch. What about Marcel's boss, Colonel William H. "Butch" Blanchard? What did he have to say about it?

According to one of "Butch" Blanchard's longtime friends, the late Brigadier General Woodrow "Woodie" Swancutt, who had flown B-29s under Blanchard at the RAAF in 1947, "It was supposed to have been one of the first UFOs in hand. At first Blanchard thought he had something, but after considerable communication between him and [General] Ramey . . . Colonel Blanchard seemed quite content to accept [the weather balloon explanation]—as long as it was out of his hands, anyway."[1]

We know from Blanchard's first wife, Ethel Simms, that her former husband knew that the wreckage he had sent to Fort Worth was no weather balloon. "At first he thought it might be a Russian device because of the strange symbols on it," she said. "Later on, he realized it wasn't Russian either."[2] However, when pressed by his family at the dinner table for an answer on a number of occasions, according to his daughter Dale, he would stare off into space as if in a trance and repeat over and over again, "The Russians have things you wouldn't believe."[3]

Art McQuiddy was, in July of 1947, the editor of Roswell's second newspaper, the *Roswell Morning Dispatch*. He was also a good friend of Colonel Blanchard's, and the two would often get together for "off the record" discussions. For several months after the incident, McQuiddy had tried unsuccessfully to get Blanchard to tell him the real story. "He repeatedly refused to talk about it," McQuiddy said. Then, in October or November of 1947, McQuiddy tried again, "... when we were a bit more relaxed than usual. Blanchard reluctantly admitted [to me] that he had authorized the press release. Then, as best I can remember it, he said, '*I will tell you this and nothing more. The stuff I saw, I've never seen anyplace else in my life.*' That was all he would say, and he never told me anything else about the matter."[4]

The former mayor of Roswell, William Brainerd, was more succinct when he told a similar story of the time when Colonel Blanchard returned to Roswell a few years after the event, and he found himself in Blanchard's presence at an official function. Later the same day, he was sitting across from Blanchard at dinner. During the course of the meal, he finally asked Blanchard the obligatory question about the 1947 incident. "[It was] the damnedest thing I ever saw," was Blanchard's only comment.[5]

Chester W. "Chet" Lytle Sr., was an old friend of ours. He passed away in 2004 at the age of ninety-two, having been the nation's oldest living hemophiliac. Having founded and owned his own leading-edge communications company in Albuquerque, he was contracted with and involved in many top secret government projects, including the famous Manhattan Project during World War II that developed the first atomic bomb. Another friend of ours is the well-known UFO researcher, lecturer, and author, Robert Hastings. In 1998, Hastings was researching a book when he requested an interview with the normally reluctant Chet Lytle, but this time the eighty-six-year-old Lytle agreed. According to Hastings, at some point during the interview, the subject turned to Roswell. "Unexpectedly, Lytle told me that he had heard from a highly credible source that the object recovered near Roswell was indeed a crashed extraterrestrial spacecraft. That source turned out to be none other than William H. Blanchard, the base commander of Roswell Army Air Field at the time of the incident." Lytle related to Hastings that in February 1953 he and Blanchard, who was by then a general, had been visiting an air base in Alaska, General Blanchard as the Eighth Air Force's Deputy Director of Operations, and

Chester W. Lytle Sr. in a 1998 photo. (Photo courtesy of Tom Carey.)

Lytle as a member of the Atomic Energy Commission. Upon the completion of their missions, Lytle was anxious to get home ASAP, as his wife was about give birth to a son. Hearing this, Blanchard, a "very close friend," offered to personally fly Lytle to the Air Force base that was nearest Lytle's home in Chicago in a bomber. During the long flight, the subject of UFOs came up between the men. The Washington, D.C. sightings had occurred only the previous July, and there had been some recent sightings in Alaska. Suddenly, the general mentioned Roswell. He then told a somewhat startled Lytle that an alien spacecraft with four dead humanoid beings aboard had been recovered in July 1947. Hastings, who knew about the Roswell Incident but had never heard that before, wanted to make sure he had heard Lytle correctly. So, he turned Blanchard's statements about Roswell into a question for Lytle, to which he unhesitatingly replied, "Oh, absolutely!"

Additionally, according to Hastings, Chet Lytle said that he had heard from another high-level Air Force source that some of the alien bodies had first gone to Muroc Field (now Edwards Air Force Base) in California before eventually going to Wright Field in Dayton, Ohio. He also said he heard about autopsies carried out on the bodies from people who were involved, and that the recovered alien craft was stored at Wright—in Hangar #5, not in the infamous Hangar #18! "I had the highest clearances . . . but I was never allowed into that area."

Three years later, Hastings contacted Lytle again to try to arrange a follow-up interview, as he had additional questions to ask him. Lytle reluctantly agreed to meet with Hastings again, but this time he was not talking. "After you and I met, I was paid a visit by someone who wanted to know why I was talking to you and Don Schmitt." An awkward silence passed between them, and Lytle finished the meeting with, "I've spent a great many years developing my business here, and I don't want to take any chances." Hastings then left and never saw Chet Lytle again.[6]

As for Blanchard's boss, General Roger Ramey, commanding officer of the Eighth Air Force in Fort Worth, we know from public records that Ramey was the architect of the weather balloon cover story that chilled press interest in the story for three decades. But what about in less public settings? What did he say about it then?

We located a firsthand witness several years ago who had been stationed at Fort Worth AAF back in 1948. As an enlisted crewman on a B-29, one day he

was waiting on the tarmac to board his aircraft along with the rest of the crew. Also waiting with the crew on this day was General Ramey himself. One of the officers in the crew was overheard to ask Ramey about the 1947 Roswell events: "What about it, General? What was that stuff?" To which Ramey responded, "It was the biggest lie I ever had to tell. . . . [It was] out of this world, Son, out of this world."[7]

The late stage, screen, and television actress-turned-Vietnam-War-photo-journalist Elaine Shepard was married to an Air Force general. Their circle of friends included the General and Mrs. Ramey. According to her, "Latane [Mrs. Ramey] told me that before her husband died [in

Brigadier General Roger M. Ramey, c. 1947. (Photo courtesy of the Army Air Forces.)

1963], he told her that the 1947 Roswell crash involved a 'spaceship' and not a weather balloon, as he had previously stated publicly."[8] Shepard would not divulge any additional details of her conversation with Mrs. Ramey because she wanted to include it in a forthcoming book. The book never happened, as Shepard passed away soon thereafter.

Retired General Arthur E. Exon was a lower-ranked officer in 1947 when he was stationed at Wright Field in Dayton, Ohio. At the time of the incident, he was a student there participating in a two-year industrial administration course at the Air Force Institute of Technology (AFIT). A year later, he was assigned for the next three years to the Air Material Command Headquarters at Wright, where the Roswell artifacts had been sent after the recovery. Later (August 1964 to January 1966), as the base commander at Wright-Patterson Air Force Base, Exon was in a position to know things; even if he did not have firsthand access to the Roswell artifacts, he knew people who did. According to Exon, after conducting a series of metallurgical tests upon the Roswell wreckage, the overall consensus among the scientists involved in the testing was that "... the pieces [of wreckage] were from space."[9]

21

"THE PIECES WERE FROM SPACE"

From the moment that a craft of unknown origin descended tragically from the summer skies and crashed in the high desert of New Mexico in 1947, high-level officials outside of Roswell clamored to get the materials into their hands for analysis.

At that time, and certainly as a consequence of World War II, there were a large number of national laboratories that specialized in a wide variety of military technologies. But there was only one facility that was dedicated to the science of reverse engineering: the FTD.

The Foreign Technology Division (FTD) at Wright Field in Dayton, Ohio, was a key intelligence organization responsible for the breakdown and analysis of all weapons and equipment captured during the war. When the military got their hands on something of foreign design—or of an even higher level—the captured technology would go to the FTD for dissection.[1] As we have detailed throughout this book, Wright Field was Major Jesse Marcel's original destination when he made his first trip escorting wreckage from Roswell Army Air Field.

(NOTE: For the historical record, the actual Foreign Technology Division wasn't established until 1961. Before then, it was known as the Air Technical Intelligence Center [ATIC], and back in 1947 it was known as "T2 Intelligence," which was attached to the Air Material Command [AMC] headquartered at Wright Field under the command of General Nathan F. Twining. FTD has become the more common affiliation and, for continuity, will be the referent used throughout this chapter.)

Instead, that flight was diverted after takeoff to Fort Worth, Texas, where General Roger Ramey announced to the press that what Major Marcel had recovered was, in fact, the remains of a weather balloon. Officially, the flight scheduled to carry Marcel to Wright Field was cancelled, and he was ordered, after a one-night layover, to return to Roswell AAF. But an official FBI tele-type message from its Dallas office to Director J. Edgar Hoover disputed that version of events, alleging that a clandestine flight carried the recovered Roswell material to Wright Field as originally planned, and that the FTD had made preparations for its arrival while the Army was still in the midst of spinning its weather balloon story to the media.[2]

Brigadier General Arthur E. Exon described to us what happened after the flight reached the FTD. Exon, then a lieutenant colonel, was an adminis-tration student in technology at the FTD. "We heard the material was coming to Wright Field," he said. Analysis of the debris was performed in the FTD's various labs: "Everything from chemical analysis, stress tests, compression tests, flexing. It was brought into our material-evaluation lab. I don't know how it arrived, but the boys who tested it said it was very unusual." Exon also described the material: "[Some of it] could be easily ripped or changed," he said, but did not elaborate. "There were other parts of it that were very thin but awfully strong and couldn't be dented with heavy hammers. It was flexible to a degree." According to Exon, "Some of it was flimsy and was tougher than hell, and the [rest] was almost like foil but strong, It had [the FTD analysts] pretty puzzled."[3]

Retired Brigadier General Arthur Exon in a 2001 photo. (Photo cour-tesy of Tom Carey.)

The lab chiefs at Wright Field set up a "special project" for the testing of the material. "They knew they had something new in their hands," continued Exon. "The metal and material was unknown to anyone I talked to. Whatever they found, I never heard what the results were. A couple of guys thought it might be Rus-sian, but the overall consensus was that the pieces were from space. Everyone from the White House on down knew that what we had found was not of this world within twenty-four hours of our finding it." When asked what he thought about the components' physical makeup, he said, ". . . I don't know, at that time, if it was titanium or some other metal, or if it was something they knew about and the processing was something different."[4]

General Exon's experience with the recovered Roswell remnants wasn't limited to the work at Wright Field either. A number of months later, he told us, he flew over central New Mexico and checked out the crash site he had heard about while stationed back East. "[It was] probably part of the same accident," he said, "but [there were] two distinct sites. [At] the northwest [site], pieces found on the [Foster] ranch, those pieces were mostly metal." The general also confirmed having seen the gouge that others had reported. Exon said, "I remember auto tracks leading to the pivotal sites and obvious gouges in the terrain."[5]

When asked about the bodies, he said, "I know people that were involved in photographing some of the residue from the New Mexico affair near Roswell. There was another location where . . . apparently the main body of the spacecraft was . . . where they did say there were bodies." Asked if the bodies had been sent to Wright Field, Exon said simply, "That's my information . . . people I have known were involved with that."[6]

General Exon also commented about specific information originating in Roswell: "Blanchard's leave was a screen. It was his duty to go to the site and make a determination. Blanchard couldn't have cared less about a weather balloon."[7] This bit of inside information was also mentioned by Lieutenant Colonel Joseph Briley, who was assigned as the operations officer at headquarters at Roswell in 1947. "Blanchard's leave was a blind," mused Briley. "He was actually setting up a base of operation at the crash site north of town."[8] We have also been told by airmen and NCOs who were there that Blanchard clandestinely moved his office to one of the enlisted men's barracks on the base to get away from the press. He may also have checked in at the base guardhouse (a.k.a. "the brig") for a time—the last place anyone would expect to find the base commander spending his "leave."

Exon elaborated, "I know that at the time the sightings happened, it was [up] to General Ramey . . . and he, along with the people at Roswell, decided to change the story while they got their act together and got the information into the Pentagon and into the president."[9]

Because of how publicly outspoken Exon, a high-ranking officer, was about the incident, we anticipated the reaction in Washington. During the GAO investigation of Roswell in 1994–95, Exon was interviewed by a number of high-level congressional staff members. One of the discussions took place at his home on December 2, 1994. Exon was extremely guarded in these talks, and one of the staff members entered into his report, "General Exon is afraid. He was afraid he was being monitored at that point. He was probably afraid his whole house was bugged."[10]

It is ironic to point out that two of the principal characters in the entire Roswell saga went on to play intricate parts in the U.S. government's twenty-two-year (1947–1969) official investigation of the UFO phenomenon—namely, General Roger Ramey and the FTD. Ramey, for all of his stoic efforts to discredit all of the flying disc reports at that time, would become a consultant to the Air Force's Project Blue Book. And in 1952 he was called their "saucer man" and one of their top "UFO experts." Ramey played an important role in debunking the subject and conducting damage control for the Air Force throughout the mid-1950s. We contend that they could always use a good weather balloon to pull out whenever a case got too hot.[11]

And, most curious of all, the FTD supported Project Blue Book with its resources and was responsible for conducting analysis work on what the Air Force investigators gathered. In fact, AFR 80-17, an Air Force regulation issued by the U.S. Department of the Air Force in Washington, D.C., was the official form used by Project Blue Book! The document also states "If the final report is deemed significant, the FTD will send the report of its findings to AF Systems Command, Andrews AFB, Washington, D.C., which will send a report to Headquarters USAF."[12]

If Roswell represented nothing more than a misidentified weather balloon—the Air Force's contention for the first forty-seven years in the history of the case—why would some of our country's "best and brightest" head for the little town of Roswell, located "south of lost," as the sarcasm dripped in those days, even before the dust had cleared? To view the remains of a weather balloon?

In the summer of 1947, Earl Zimmerman was a re-enlistee working at the RAAF base radio shack as a high-speed code transmission radio operator. During off-duty hours, he moonlighted as a bartender at the base Officers' Club. He heard the many rumors about the crash of a flying saucer while working at the club and around the base, including something about investigating the discovery of one under the guise of a plane crash (also see account of Lieutenant Chester Barton in Chapter 11). Zimmerman remembered, at the time of the UFO crash in 1947, seeing General Ramey, the commanding officer of the Eighth Air Force in Fort Worth, Texas, to which the 509th Bomb Group in Roswell was attached, in the RAAF Officers' Club on more than one occasion. On several of these occasions, he recalled seeing General Ramey with none other than one of our nation's most beloved national heroes, aviator Charles Lindbergh. "I heard that they were on the base because of the flying saucer business. There was no publicity about Lindbergh's visits, and I was very surprised to see him in the club. I think he came to Roswell with Ramey . . . from Puerto Rico."[13] Speculating here, it appears that the well-known "Lucky Lindy" was first flown to Puerto Rico (a frequent stopover point for Eighth Air Force bombers), where he was picked up by General Ramey and surreptitiously flown into Roswell to offer his opinions. Was all this James Bond chicanery and intrigue really over a *weather balloon*? And, if the true wreckage were indeed the weather balloon seen in photographs on the floor of General Ramey's office in Forth Worth, why were they in Roswell?

The RAAF Officers' Club was doing a brisk business at the time of the 1947 UFO crash. In addition to the normal complement of Silverplate flight officers seeking to unwind, gripe, or otherwise drown their sorrows in the bubbly, a number of new faces all of a sudden turned up at the club. In one instance when Earl Zimmerman was moonlighting at the OC, he noticed an officer who was unknown to him. Considering the rumors and overall tenseness on the base at the time, he called the officer's presence to the attention of Colonel Blanchard, who apparently was spending his "leave" very close to home. According to Zimmerman, Blanchard did not know who this person was either, so he went over to him to find out. Blanchard later told Zimmerman that there was "no problem." The man turned out to be a CIC agent from out of town.[14] Something was definitely up, and the influx of "suits" from Washington, Los Alamos, White Sands, and Dayton, Ohio, seen around the base was proof that something unusually important was taking place.

Besides Lindbergh and the CIC agent, one of the other new faces in the RAAF Officers' Club was that of Ben Games. In World War II, Korea, and Vietnam, he flew bombers and night fighters, recording 737 hours of flying combat in the process. He was awarded the DFC (Distinguished Flying Cross) and the Bronze Star, among many other medals and commendations. He also worked on some of the very first flight simulators ever developed. In July of 1947, Games was a young Army Air Forces lieutenant on flight status as the personal pilot of Major General Laurence C. Craige, chief of the Army Air Force's Research and Engineering Division under General Curtis LeMay, who was then the Air Force's Chief of R&D (research and development) at the Pentagon. In October of that same year, Craige would assume LeMay's job when LeMay was selected to command the U.S. Air Forces in Europe headquartered in Wiesbaden, Germany. Craige was an MIT graduate who was an acknowledged technical and military genius. In 1942, he was the first American military pilot to fly a jet-propelled plane (the Bell XP-59 Airacomet). According to Games, he flew General Craige on an unscheduled flight to Roswell Army Air Field from Wright Field directly after the (UFO) crash was discovered. On the flight, Games recalled that Craige didn't talk much and was very tight-lipped about the nature of the visit and the reason for the flight. Upon landing, Craige was picked up and whisked away in Colonel Blanchard's staff car, while Games, who had not been told the reason for the flight, was left standing there. So, like any red-blooded pilot with time on his hands, he headed for the Officers' Club to contemplate his future. Craige was gone for a number of hours, and it was during his wait at the Officers' Club that Games overheard much excited talk about the crash of a *flying saucer* with "little bodies" not far from Roswell. Craige finally returned from wherever he had been and ordered Games to fly him directly to Washington to meet with President Truman. Games believes that General LeMay, who was a close friend and professional associate of General Craige, likely helped brief Truman on the Roswell matter. Games said that he is certain that there is no record of Craige's meeting with Truman after his trip to Roswell. He said that this was not unusual, because he was personally aware of many meetings at the White House for which there are no records—yet they took place, because he flew the generals there! He also said that General Craige himself was an intense and secretive man, and that records of his schedules and activities are largely nonexistent. Games also knew General LeMay well, and said that LeMay likely was aware of everything about the crash and was perfectly suited to the task of dealing with an event such as this. Just a few months after Roswell, it was General Craige, now the head of USAF R&D, who ordered the establishment of the Air Force's first official study of UFOs, Project Sign. Soon thereafter, in military time, Craige received his third star and promotion to lieutenant general. He would go on to other commands throughout the world, including commanding the Allied Air Forces in Southern Europe in Naples, Italy. He retired from the Air Force in 1955 and passed away in 1994.

It wasn't until 2003 that Games became aware that the Roswell event was such a "public phenomenon." He explained that he was living in many countries throughout many years when he headed a small airline company, and it wasn't until much later that he realized Roswell was a big thing, and how his role in the event fit. Waxing philosophical, Ben Games today believes that what came down at Roswell was not made by us and was not a balloon. He believes that it was extraterrestrial, based upon what he knows about his former boss, Laurence Craige. "There is simply no way whatsoever that Craige would have gotten involved if what had crashed was something that we flew or that we already knew about. He thought fifteen years ahead all the time and was more aware than anyone else of even the most secret experimental aircraft. And why did Craige go see (President) Truman directly after Roswell, if what had happened wasn't of the utmost technological or national security importance?" Regarding the Air Force's current explanation for the Roswell crash—that it was a balloon train from a top secret project, Project Mogul—Games laughs, "No matter how large the balloon, or how many balloons were put into the train, any pilot or trained officer would know that it was still a balloon!" Using his insider's knowledge, Games conjectures that the Roswell crash debris was sent to different locations for study, and that it was all very compartmentalized. He believes that a very small group of people at the time of the crash knew the full story, and that General Craige and General LeMay were among this small group.[15]

With all of this going on, the obvious conclusion to be drawn is that both Ramey and the FTD were completely aware of the true nature of the Roswell material. Until they could amass all the intelligence necessary to provide the complete picture from a national security standpoint, the situation would remain fluid and the search for answers would continue. Meanwhile, total secrecy was the only means by which they retained control over the predicament. Startling as it would seem, after seventy-five years, the scenario hasn't changed much—four official explanations for Roswell notwithstanding. As were all the others who had been in Roswell at the time of the recovery, at Fort Worth when Marcel brought the material to the office of General Ramey, or at Wright Field to marvel and shudder at the possibilities, Arthur Exon was convinced that the wreckage had come from something not manufactured on this Earth.

22

"IS *THAT* WHERE THAT FLYING SAUCER CRASHED IN 1947?"

The year was 1987, just three months past the fortieth anniversary of our heralded case in point. The Berlitz/Moore *Roswell Incident* book made a slight blip outside the UFO community, but we are sure that, much to the relief of the Pentagon, the press paid absolutely no attention. Not even the city of Roswell batted an eye. But, as we will explain, the powers that be hadn't received that memo. Considering the level of ongoing concern four decades after that big debacle over a mere weather balloon, we had to wonder, if it were truly something so mundane, why would anyone still care?

Ranch hand James Parker had just dropped off his children at the bus pickup to the Corona Elementary School. Once they were safely aboard, he left a dusty trail in his rearview mirror as he headed back to the ranch by steering his Chevy truck up the Hines Draw to start his daily routine of chores. Work would be delayed this day, however, for no sooner had he passed the old Hines House than, up ahead of him a quarter of a mile or so, Parker saw an unknown vehicle lumbering down a seldom-used two-track dirt road toward him. Parker related that it was a pickup hauling a small trailer with two occupants who must have been just as shocked to discover that they too were not alone. Almost losing control as they reached a drainage ditch, they still managed to maneuver the fast-moving truck into making a wide-sweeping right turn into the draw in an attempt to try to elude the now pursuing rancher. To Parker, a trespasser was a trespasser, but the situation was quickly escalating into a scene from *Walker, Texas Ranger* as both vehicles were exceeding speeds of fifty miles per hour. On dirt roads normally meant for maximum speeds of half that, there were only two possible outcomes. Fortunately for all involved, cooler heads prevailed. The chase scene would continue for a few more miles when, much to the relief of the angry worker, the pursued finally began to slow and come to a stop. Parker abruptly pulled up right behind the waiting vehicle and, with Winchester in hand, leaped from his pickup and lunged forward towards the driver. "What the hell do you think you're doing? This is private property," blasted the highly agitated Parker.[1] What he would hear next (which we will reveal at the end of the chapter) was a page right out of *The X-Files*.

This was far from the first time intruders ventured onto this remote parcel of land, which heretofore attracted only grazing cattle, sheep, and rattlesnakes[2]—not what you would call a tourist attraction by any stretch of the imagination. But the incident back in 1947 changed all that. Something beyond human comprehension took place, and the military took measures that were beyond any standard salvage operation—even for a top secret Mogul balloon. And again, for the readers' edification, we cannot emphasize enough times—not a single, structural piece of Mogul was classified in any manner, shape, or form. Prosaic items such as rubber, tinfoil, balsa wood, tape, and twine, no matter how you cut them, are *still* rubber, tinfoil, balsa wood, tape, and twine!

The late Major Charles A. McGee would occasionally be assigned sentry duty at one of the gates at the RAAF, starting in the late summer of 1947.[3] The main gate entrance went north up Main Street, taking you into downtown Roswell, whereas the east gate took you from the B-29 hangars (Hangar P-3 being one them) to the freight yard just outside the base, and the south gate took you more than a mile from the inner perimeter fence to the base radio transmitters and antenna towers. The major was not assigned to any of these secured posts but rather to the west gate, which basically led to nowhere. But, according to McGee, the main purpose of this exit was for base personnel to secretly bypass the city of Roswell and head north using the back roads—to the debris field.[4]

The west gate on the old RAAF base is still there but blocked now to traffic. Don Schmitt looks out to the northwest where a back road was used to travel back and forth to the 1947 crash sites, thus avoiding the town of Roswell. (Photo courtesy of Tom Carey.)

The main "back road" to the debris field was the Pine Lodge Road located at the north end of Roswell where it intersects Highway 285. According to Rogene Cordes, whose parents owned a large ranch north of town, she and her family were barred from using the Pine Lodge Road by camouflaged men with machine guns and were thus delayed in getting to their ranch the weekend of July 12 to 13. She also recalled that she overheard talk at the bank where she worked from "early rising ranchers" about long trucks covered in canvas going to the base before dawn at the time of the crash.[5]

"Typically, after heavy rains up in the Lincoln County region, that's when they'd make their move," McGee said. Men would slip out from this unoccupied area of the base proper and rendezvous with accomplices pulling horse trailers. Then they'd change into ranch attire and inconspicuously ride through the wreckage site on horseback, scouring the ground for any missed remnants from the crash.[6]

We are once again asked to accept that officers from the one and only atomic base in the world at the time would drive two hours each way to go play cowboys out in the middle of the desert looking for balloon parts. And, as McGee still expressed his amazement to us more than fifty years after this clandestine activity, "This went on for about two and a half years!"[7] Adding further testimony to this unusual story was none other than Walter Haut, who stated, "They were still searching the desert for a couple of years after the crash."[8] Clearly, Bill Brazel's shocking experience two *years* after the original,

213

full-scale recovery operation demonstrates the ongoing concern the military has displayed in retrieving any evidence contrary to the official balloon explanation—even if it should take years!

It was due specifically to the assistance of Bill Brazel that the J. Allen Hynek Center for UFO Studies was able to conduct the very first archaeological dig at the site in September 1989. This was just five months after Brazel had first driven us to the precise location where his father, other witnesses, and he had discovered and retrieved certain "features upon the landscape" that were out of the ordinary.[9]

But before proceeding with this historic project, there was one major obstacle to overcome: the parcel of land in question was and still is owned by the Bureau of Land Management (BLM), which happens to be administered by the U.S. Department of the Interior.[10] The objective of the BLM is to protect and conserve land, water, and natural resources. To begin the Hynek Dig, special permits were required with strict land-use limitations. During the course of the four-day expedition, vehicles were allowed to travel in and out to the site on only one two-track road. Next, the vehicles had to be parked, and all other travel had to be done on foot. Trenching and on-site excavation were approved, but only at a minimum, and, as a final requirement, the team was to leave the area environmentally sound with little trace of human impact.[11]

As a convoy of four vehicles departed their excavation at the debris field for the last time on a late Sunday afternoon, they finally reached the asphalt road that would take them into Corona. But no sooner had they turned onto the pavement than they noticed, parked on the shoulder up ahead, an open jeep and a man in fatigues talking on a two-way radio. Strange as it seemed, everyone passed it off as just coincidence. These feelings would quickly turn to uneasiness, for, as the group continued down the highway, up ahead was another vehicle and a man talking on a handheld radio. Someone in the party suggested that they might have been BLM officials, which drew an immediate retort that such government employees would be neither driving jeeps nor wearing military fatigues. The group attitude remained that they had done nothing wrong, and they should continue behaving that way. But with each new mile they crossed towards the little town of Corona, anxiety was gradually replaced by relief. That feeling was to be short-lived, however. As they rounded the next curve, there—to everyone's shock—was a *third* jeep reprising the exact same scenario as the previous two. Within moments, the CUFOS scientific director, Dr. Mark Rodeghier, turned to one of the coauthors (Don Schmitt) and half-jokingly said, "I think we're in trouble." Still, the train of vehicles wound nervously into the sleepy town of Corona and pulled into the only filling station open for business. All eyes, however, remained riveted down the road just traveled to see if anyone had followed, and, as though out of nowhere, a phantom jeep carrying two men in similar fatigues pulled up alongside one of the refueling cars. "We're strangers in town and looking for a good place to go deer hunting. Can you help us?" called out the driver. One from the group apologetically offered the same excuse and suggested they ask the station manager. With that, the CUFOS

As seen in a 2010 photo, we arrive at the impact site. (Photo courtesy of Tom Carey.)

Sharing the moment of discovery of the impact site are [L to R] BLM archaeologist Pat Flanary, his wife Gretchen, Don Schmitt, and the late Roswell photographer Jack Rodden, as seen in a 2010 photo. (Photo courtesy of Tom Carey.)

team then made a quick beeline for the next road to Albuquerque and never looked back. The unexpected "guests" were not seen or heard from again—at least not by anyone from the archaeological project.

A few weeks would pass when the ranch supervisor, Jeff Wells, called with some disturbing yet amusing news. In a half serious tone, Wells said, "Someone was here today from the BLM, and they're pretty upset. I know you guys left that pasture in perfect condition. I know, because I went out there right after you left and checked for myself. Well, we were just out there again, and someone's trying to get you all in a lot of trouble." According to the BLM, the entire site was ripped up by someone who had crisscrossed it from one end to the other and had also spun circles over a large area of the grazing land. It was also quite evident that more than one vehicle was involved. Fortunately, the ranch foreman could vouch for the innocence of the archaeological team. If not, it would be a safe bet that all future plans for returning to continue work at the debris field would have been rejected. An even safer bet would be to assume that *that* was precisely someone else's intent. And isn't it funny that the only vandalism that occurred was at *that* historic location? And how could they know exactly where to go to perform their setup job? Maybe it was because they had been out there many times before through the years. Maybe it was because they too are still searching the ground for leftover physical evidence, all the while confidently relying on their fallback explanation, like the intrepid detective Frank Drebin: "Nothing to see here!"

As James Parker approached his trespassers with no prior knowledge of the preceding events, the words "U.S. Air Force" jumped from their pickup's door like a flashing neon sign. Inside, with numerous topographical maps scattered on the dash and console between them, sat two middle-aged men. The driver, a corporal, and his PFC partner looked very out of place due to their ages. This is a common ploy by military officers to avoid identification when conducting undercover field work. Nevertheless, they were still on private property. "Get the hell off this land," ordered Parker. Neither of the two interlopers identified themselves. The driver nervously tried to explain the awkward situation by offering the lame excuse that they were simply there to check out the final location of a low-altitude radar tracking facility for monitoring drug smugglers coming in from Mexico. "Say, what? Can you please repeat that, Mister?" The rancher had never heard of any such arrangement from his boss—and neither had anyone else, for that matter. In point of fact, there is no such tracking system anywhere in New Mexico. Everything else being equal, pointed rifle barrels usually end discussions, and the Air Force was forced to retreat—at least that time—to fight another day. But before driving off into the sunset to abandon their so-called assignment, the driver posed this illuminating query, "By the way, you saw where we were originally coming from. Is *that* where that 'flying saucer' crashed back in 1947?" Parker's mouth dropped open. Unlike the military, which has an ongoing concern for the 1947 incident, he had no idea what the man was talking about.[12]

23

DEATHBED CONFESSIONS: "I DIDN'T SEE THE WHITE LIGHT, BUT I DID SEE THE ALIENS AT WILFORD HALL"

In courts of law, so-called deathbed confessions are accorded special weight and consideration because of the belief that when a person knows that death is pending, that person will want, in the end, to have his or her conscience cleared and leave truth as a lasting legacy. Perhaps the most significant deathbed testimony has been that of the former provost marshal at the Roswell Base in 1947, Major Edwin Easley. When first interviewed by Roswell investigator Kevin Randle, all Easley would say was that he couldn't discuss the Roswell Incident; that he was still sworn to secrecy. Over and over, Easley would repeat that same phrase to each question that Randle asked.[1] But sometime thereafter, he confirmed to his two daughters and granddaughter his participation. When he lay dying at Parkland Hospital in Dallas, Texas, in June of 1992, his granddaughter brought him a gift while her mother and aunt kept vigil at their father's bedside. She held it up to her grandfather's eyes in the hopes that he would appreciate her thoughtfulness. With a look of total astonishment, he turned from the outreached arms of the young girl and sighed the words, "Ohhh . . . the creatures!"[2] What was the gift, you ask, that brought such an intense reaction from an intrepid man who had maintained his silence for forty-five years? It was simply a copy of the first Randle and Schmitt book, *UFO Crash at Roswell.*[3]

Easley wasn't the only one to make such a confession. Just before he passed away in November of 1995, the former Roswell base adjutant in 1947, Major Patrick Saunders, wrote on numerous copies of the paperback edition of the 1994 Kevin Randle and Don Schmitt book *The Truth about the UFO Crash at Roswell*, which he sent to family and close friends, this cathartic statement: *Here's the truth, and I still haven't told anybody anything!* (See Appendix III.) Several months before he died, he confided to other close friends that they were faced with a technology greater than ours, and that "We had no idea what *their* intentions might be."[4]

For years, we had assumed that Saunders had written his note on that page out of convenience because it was the first page inside the cover of the book. Years later, after we were able to determine that the book's key witness, Frank Kaufmann, was a fabricator of documents as well as a purveyor of misinformation (and therefore impeached as a reliable witness to the 1947 Roswell events), Kevin Randle and Tom Carey, independently of each other, reread the Saunders note, especially the passage in the book it was sitting atop, and came to the same epiphany: that the "truth" Saunders was referring to was not the entire book, but only the passage just under his handwritten note. In a January 2007 email to Carey, Randle shared his revelation:

> [Saunders] didn't need to cram his words at the top of this page unless there was important information on it. Saunders confirmed, from his point of view, that this was the truth. He might not have agreed with everything in the book, but he did with this specific page. . . . In a 1997 letter to me, one of Saunders' children wrote, "At one point, he bragged to me about how well he had covered the 'paper trail' associated with the clean-up!" . . . I skipped over this when I read the letter ten years ago. When I saw it this time, it sent chills down my spine. Here for the first time was information that suggested a cover-up in place in Roswell . . . beyond what we can deduce from what others have said.

Very well, indeed! In 1993, at the request of New Mexico congressman Steven Schiff, after Schiff had gotten the runaround from Secretary Dennis DeConcini and the Department of Defense for his questions about Roswell, the Government Accounting Office (GAO) was tasked by Schiff to locate documents pertaining to the 1947 Roswell events—not to prove or disprove the Roswell case, but to determine if the documents had been properly handled and classified. At the conclusion of the GAO's investigation in 1995, Schiff reported that the GAO was unable to locate any Roswell-related documents (beyond the very few that were already known to researchers), that such pertinent documents—if they ever existed—had been destroyed without apparent authority.[5] Thanks to Patrick Saunders, we now know at least one of those "authorities" responsible for the disappearance of the Roswell paper trail. We also now know that there was a cleanup going on, not only out at the crash sites, but also back at base headquarters.

As participants in the Roswell events of 1947 expire at an ever-increasing rate, it should be expected that we encounter more confessions of the "deathbed" variety as time passes, and such is indeed the case. Sarah Mounce, whose husband, Private Francis "Frank" Cassidy, was an MP in the 1395th Military Police Company at Roswell in 1947, relayed to us the information that her husband, during his final days in 1976, confessed to guarding Hangar P-3 and seeing the bodies inside;[6] another woman, Wanda Lida, told us that her husband, Corporal Robert J. Lida, during the last remaining months of his life in 1995, finally told her of his involvement in the Roswell events of 1947. After seeing a program on TV that featured the Roswell Incident, she at last asked him, "Well, Dear, is it true?" He answered, "Well, I suppose it's time I should tell you. I've been meaning to for a long time. It's true."

Corporal Lida had been an MP with the 1395th at Roswell in 1947 and confirmed to Wanda that he was simply "grabbed" one day and told to report to Hangar #3. He was given a gun and told to stand guard at the hangar with other similarly marshaled base personnel. While on guard duty there, he waited for an opportunity to look inside the building. Lida swore to his wife that he observed wreckage scattered about inside and a number of "small bodies" being prepared for shipment elsewhere. Asked if she believed her husband when he told her this, Wanda replied without hesitation or reservation, "Absolutely! He was telling me the truth when he knew he didn't have much longer."[7]

Sergeant Homer G. Rowlette Jr. was a member of the 603rd Air Engineering Squadron at the RAAF in 1947. He was career military and retired as an NCO after twenty-six years of dedicated service to his country. Before passing away in March of 1988, he finally conveyed to his son Larry the following startling information about his involvement with the "crash of the flying saucer." Homer was part of a cleanup detail sent to the impact site north of Roswell. Larry was told that his father had seen everything. He had handled the "memory material," which, according to Homer, was "thin foil that kept its shape." If that weren't enough, he described the actual ship, which was "somewhat circular." But what followed caught his son completely by surprise: "I saw three little people. They had large heads and at least one was alive!" His father ended by adding that there were *three* sites—the one just north of Roswell, and two others near Corona.[8]

According to Larry's sister Carlene Green, "My father was a very honest, honorable, and trustworthy man. He never lied to me." Her father had never mentioned his tenure at Roswell to her until just two weeks before he died. Homer was given only days to live. Carlene, who at the time knew nothing about the story her dad had secretly passed on to her brother, was nervously waiting with her father as he lay on a gurney in the hospital about to be wheeled into the operating room. Still totally lucid, he painfully motioned for her to come close so he could speak, in case he didn't

have another chance, which was clearly the impression his daughter had at the time. She had no idea he would confess the following: "I was at Roswell when they recovered the spaceship in 1947. I was involved. I saw it. It's all true." Homer told Carlene that he was sorry for never saying anything before, but he was told to "keep quiet, or else!"[9]

Since the publication of the first edition of this book, corroboration for Homer Rowlette's story has come from two sources, both former members of Rowlette's 603rd Air Engineering

RAAF Sergeant Homer Rowlette.
(Photo courtesy of Don Schmitt.)

Squadron at the RAAF. Master Sergeant Harry Telesco was already gone when we located and contacted his family in 2007. According to his daughter, her late father had spoken many years ago of being at the impact site and seeing the wreckage and little bodies there. He also told of seeing one that was still alive. Telesco's daughter said that, upon hearing this, the family proceeded to gave him a "rough time" about it, and he never spoke about it again.[10] A similar account was given by former 603rd PFC James Saine. According to his surviving family members, he was also at the impact site, where he witnessed what was left of the craft, as well as the bodies.[11]

One of the earliest examples of a deathbed confession came from the late Sergeant Melvin E. Brown, who was with K Squadron back at the time of the incident. Brown would take the historic first manned landing on the moon in July of 1969 as an impetus to tell his family the truth about Roswell. Unfortunately for Brown, they were reluctant to believe him. Still, his wife and two daughters remembered his stern warning not to tell anyone else, because "Daddy will get into trouble."

In 1986, as Brown was on his deathbed just outside London, England, his daughter, Beverly Bean, said that her father talked about Roswell exclusively. He reiterated over and over that "it was not a damn weather balloon." Brown's wife and oldest daughter still refuse to discuss the matter. Beverly Bean, however, wanted everyone to know what her father had told her with his dying words:

> It was approaching dusk when one other soldier and I were stationed in one of the ambulance trucks at the recovery site. Everything was being loaded onto trucks, and I couldn't understand why some of the trucks had ice or something in them. I did not understand what they wanted to keep cold. Our orders were not to look under the canvas tarp in the back. The moment we had a chance, I pulled back the covering. There were bodies . . . small bodies . . . and they had big heads and slanted eyes.

His family still fears government reprisal should they say too much.[12]

After the *Unsolved Mysteries* broadcast of September 1989, a former cancer ward nurse from the St. Petersburg Hospital in Florida came forward to describe the final testimony she personally heard from one of her elderly patients. The nurse was Mary Ann Gardner, who worked at the hospital from 1976 to 1977. The patient, a woman (Gardner couldn't remember her name), had been alone in the hospital. Feeling concern for her because she had no visitors, Gardner spent as much time as she could listening to the woman's stories—especially the one about the crashed ship and the "little men" she had seen.

According to Gardner, "Basically . . . they had stumbled upon a spaceship of some kind and . . . there were bodies on the ground . . . little people with large heads and large eyes. Then the Army showed up . . . and chased them away . . . the Army people were everywhere and . . . told them that if they ever told anything about it, that the government could always find them."

The dying woman explained that she had been with a team of archaeologists and was not supposed to be there. The team had been "hunting rocks and looking for fossils . . . she had gone along with a friend." Gardner said the women was still frightened about official retaliation and soon had nothing more to say about the incident. "The woman kept looking around as if she was frightened about something and said to me, 'They said that they could always find us, so I'd better not say any more.'" Gardner then asked her, "Who? Who can find you?" The woman then answered in a low, wary voice, "The government." Within days, she expired.[13]

In the years since Mrs. Gardner first came forward with her story, and the St. Petersburg Hospital refused Don Schmitt's request for access to its records, Tom Carey decided to research the anthropological and archaeological literature in an attempt to try to identify Gardner's "dying archaeologist." He was able to identify a female archaeologist who would have been about the right age, and who had passed away in 1976 in St. Petersburg, Florida. Carey then called Mrs. Gardner several times to tell her the news and, when she didn't answer the phone, left messages as to the reason he was calling. Failing this, he then wrote a letter to Gardner containing a list of a half-dozen names of female archaeologists, including the one he had identified to see if the name itself would jog her memory. Like the phone calls, the letter was never answered (and never returned), and further phone calls as recent as 2008 were met with a gentleman's voice stating that we had the wrong number.

From all eyewitness accounts, the Roswell Incident left a devastating impact on the Chaves County sheriff, George Wilcox. Just before his widow, Inez, passed away, she related a story to her granddaughter Barbara, who told us, "She said that the event shocked him. He never wanted to be sheriff again after that. My grandmother said, 'Don't tell anybody. When the incident happened, the military police came to the jailhouse and told George and I that if we ever told anything about the incident, not only would we be killed, but our entire family would be killed!'" Barbara added:

> They called my grandfather, and someone came and told him about this incident. He went out there to the site; there was a big burned area and he saw debris. It was in the evening. There were four "space beings." Their heads were large. They wore suits like silk. One of the "little men" was alive. If she said it happened, it happened! My grandmother was a very loyal citizen of the United States, and she thought it was in the best interests of the country not to talk about it.

Inez Wilcox expired shortly thereafter at the age of ninety-three.[14]

Meyers Wahnee was a full-blooded Comanche Indian who had piloted B-24 Liberator bombers during WWII. Fondly known as "Chief" by his crewmates, Wahnee was a top-tier security officer by 1947. In July of that year, Captain Wahnee was ordered from Fort Simmons in Colorado to

Captain Meyers Wahnee. (Photo courtesy of the Army Air Forces.)

Roswell Field in New Mexico to oversee the transport of a "top secret item" from Roswell to Fort Worth via a special B-29 flight. The item in question was a single, large, wooden crate that Wahnee was to accompany in the bomb bay for the duration of the flight to Fort Worth. Apparently motivated by a featured segment about the Roswell Incident in 1980 on the popular TV show *In Search Of,* Wahnee finally broke his silence on the matter with his family during the final year of his life. According to his daughter Blanche, in a 2005 telephone interview with our investigation, her father told them:

1. The Roswell Incident was true.

2. He had flown with the alien bodies from Roswell, New Mexico, to Fort Worth, Texas (see Chapter 17).

3. There were three sites.

On his deathbed in 1982, according to Blanche Wahnee, her father gave his family one final caveat: "Whatever you do, don't believe the government. It really happened."[15]

In 2001, Dr. Roger Lier, the noted "alien implant" physician, informed us of a Roswell witness he and "Alien Hunter" Derrel Simms had been researching. They had been talking to the witness's surviving sons, but the case was requiring much more time to research—time they did not have. Dr. Lier provided us with their investigative file to continue the investigation. We quickly proceeded to interview four of Marion "Black Mac" Magruder's five sons: Mark, Merritt, Mike, and Marion Jr., who told us about their father's unexpected brush with the Roswell Incident. Honoring his security oath, he had kept it from them until he felt it was safe to tell—when he felt that it was no longer a secret—"after he saw all of the books and TV shows about Roswell in the 1980s and 1990s."

The summer of 1947 found Lieutenant Colonel Marion M. Magruder in class at the Air War College at Maxwell Field in Montgomery, Alabama. The class was filled with officers whose ranks ranged from general all the way down to lieutenant colonel, which was the lowest rank permitted. The officers chosen for the class were deemed to represent the "best and brightest" in their respective branches of the military, as well as future leaders heading into the post-war era. Magruder's class at the Air War College was scheduled to last approximately one year, from late July 1947 to early June 1948, when the officers would receive advanced training in military history, decision-making, and geopolitical strategy.

The Air War College class had been in session for a number of months when, in early 1948, all members were flown up to Wright-Patterson Air Force Base (the new USAF designation) in Dayton, Ohio. Their "opinion" was desired on a matter of utmost urgency and importance. Not knowing what to expect, the curious officers were led into a room where they were told about the recovery to Wright Field the previous summer of an extraterrestrial spaceship that had crashed near the town of Roswell, New Mexico. Most of the officers had not been aware of the crash and were startled when some of the wreckage was brought out for them to examine. The real shocker, however, was yet to come. After everyone had a chance to examine and handle the wreckage, they were taken into another room. There, they were shown something that would haunt Marion Magruder for the rest of his life.

While lying on his deathbed almost fifty years later, Black Mac recalled again his brief encounter with the "live alien" he had witnessed that one time at Wright Air Force Base with his Air War College class. The class had been told that it was a survivor from the Roswell crash. Magruder's son, Mike, had heard the story before, but this telling was meant for his granddaughter Natalie, who was there with her father. He told her that the "creature" was under 5 feet tall, "human-like" but with longer arms, larger eyes, and an oversized, hairless head for its small frame. Its other features, as described by Magruder, were similar to the descriptions of others throughout the years who have claimed to have seen the Roswell aliens: a slit for a mouth, and no nose or ears, just small orifices. In his mind, Magruder emphasized the human-like qualities of the small, childlike creature, but he told his granddaughter that there was no question that it "came from another planet."[16]

Unknown to Magruder, even on his deathbed where he passed away on his eighty-sixth birthday in 1997, the same general terms that he had used to describe the aliens were also used by two military officers who had seen them the year before. Captain Oliver "Pappy" Henderson, the pilot who had flown the first set of Roswell aliens to Wright Field on July 8, 1947, described their appearance as reminding him of the cartoon character "Casper the Ghost";[17] Major Jesse Marcel, the Roswell base intelligence officer who was dispatched to the crash site on July 6, 1947, told a subordinate that they had the appearance of "white, rubbery figures."[18] Magruder used the term "squiggly" to describe the living specimen he saw at Wright Air Force Base.[19] It would not be too much of a stretch to suggest that all three men were using their own terms to describe the same thing. According to Mike Magruder, his father later learned that the military had been conducting experiments on the alien he saw, but it had died in the process. "It was alive, but we killed it," he said. Magruder had no way of knowing that Roswell photographer Jack Rodden Sr. had told his son the very same thing many years before.[20]

Since the publication of the first edition of this book in 2007, Roswell "skeptics" have zeroed in on Lieutenant Colonel Marion "Black Mac" Magruder's story, claiming that there were no Air War College classes being held in the summer of 1947, that there was nothing to show that Magruder was ever in an AWC class, that there were no flights of AWC classes to Wright Field at the time, and nothing to indicate when Magruder might have visited Wright. Well now.

We know that Colonel Magruder was in fact in the 1947–48 AWC class from three documented sources: (1) We have a copy of the 1947–48 AWC class roster, which includes Colonel Magruder's name among the ninety attendees; (2) we have a class picture of the 1947–48 AWC class that shows Colonel Magruder standing in the first row (NOTE: the class was divided into two Groups, 1 and 2, consisting of approximately forty-five students in each section; Colonel Magruder was in Group 1); (3) our associate, Anthony Bragalia, has confirmed with the current base historian at

AIR WAR COLLEGE - CLASS 1947-48 - GROUP 1

The Air War College class of 1947-48. Lt. Col. Marion "Black Mac" Magruder stands in the exact center of the front row. (Photo courtesy of Mark Magruder.)

Maxwell Air Force Base in Montgomery, Alabama, that there indeed were AWC classes being held there in 1947, and that flights of AWC classes to Wright were taking place back then and have continued periodically up to the present time; (4) we now have a copy of Colonel Magruder's official "Marine Corps Record" (NAVMC 545A-DP) that shows all of his duty stations by year and month—it shows that he was "detached" to the Air University at Maxwell Field in Montgomery, Alabama, on July 20, 1947, but was a few weeks late in arriving, and "detached" from the Air War College there on June 7, 1948, to become an aviation instructor at the Marine Corps Schools in Quantico, Virginia. Among the entries on the form for the month of April 1948 is the following entry:

"(APR) 4-9, Temporary duty Wright Air Force Base, Dayton, Ohio."

According to Magruder's son Mark, his father never gave them a date for his visit to Wright other than to say that it was during the time he was in the

AWC class. Now we have documentation proving that Magruder was where he said he was, at the time he said he was there. The "skeptics" are correct in that the document doesn't tell us *why* he was at Wright. In their view, in order for them to accept Magruder's story, there needs to be a document that has a picture of Magruder shaking hands with the alien!

As we have seen, CIC officer Sheridan Cavitt played a major role in the entire Roswell affair. As the head of counterintelligence at Roswell, he would have been privy to most every aspect of the historic event—not that he ever admitted much to investigators. Still, his lasting legacy will be that he became the lone, star witness for the Air Force Project Mogul report. Ironically, his testimony to us and to the Air Force was a huge contradiction: from not being at Roswell at the time of the incident to being there but saying that nothing out of the ordinary took place, to going out to the ranch—not with Major Marcel, but rather with Master Sergeant Rickett?—and finding absolutely nothing, to finally going out with Marcel and recovering—you guessed it—a Mogul balloon.[21] On one occasion, his wife apologized to us when he wasn't in the room: "You have to understand," she said, "my husband is sworn to secrecy and can't tell you anything."[22] But that didn't stop us from trying. In fact, when Cavitt was terminally ill and given a very short time to live, we discussed a plan with his son Joe, an attorney, to get his father to write out a sealed statement, which could be released posthumously. Frustrated, Joe informed us each time he broached the topic with his father, and the answer was always that "he was not ready."

A C-54 Skymaster cargo plane like the one pictured was piloted by Captain Oliver "Pappy" Henderson from RAAF to Wright Field in Dayton, Ohio, on July 8, 1947, containing one or two dead alien bodies and some crash wreckage. (Photo courtesy of the 1947 RAAF Yearbook.)

Unfortunately, the intelligence officer was never ready and quietly passed away with no awards or special proclamations from the president, the CIA, the NSA, or the DOD for preserving one of America's greatest secrets. Sheridan Cavitt remains a testimony to the fact that when you know too much, you forfeit your very freedom. As his son Joe said, "It was like having a father who lived in a bubble. You literally had only half a father."[23] Apparently, the government had the nonnegotiable half. This couldn't have been more evident than what was described to us by his attending physician just days before he died. "He sat next to his bed (in the hospital) with a room full of immediate family who were reminiscing and exchanging personal stories. But not Mr. Cavitt. He just sat there not saying a word. It was though he was afraid that he would say something."[24]

Such are the control, fear, and intimidation held over these individuals, in some cases for more than seventy years. What would seriously compel them to withhold a story of this magnitude from even their immediate families? And then at the most poignant moment just prior to their death, finally break the silence and reveal what obviously weighed on them relentlessly for so many years? It is perplexing for those of us in these modern, cynical times, when the government is often held in contempt. But not for those in the "greatest generation"—no, their actions regarding Roswell throughout the past seven decades only solidifies that title. They knew how to keep their secrets. Call it devotion to duty, post-WWII patriotism, or just being a good soldier who respected the military code of honor.

Kenneth Compton was a teenager in 1970 when he saw an *unidentified flying object*. When he asked his father, who was in Air Force counterintelligence, about it, he was angrily told by him in no uncertain terms to keep his mouth shut. A few years later at Kelly AFB in San Antonio, Texas, his father suffered a heart attack, and immediate bypass surgery was ordered. After the surgery, when his father was coming to, Ken asked him, "Well, did you see the white light?" George Compton's answer to the question was startling in its matter-of-factness, "No, Son. I didn't see the white light. I saw the aliens at Wilford Hall." (NOTE: *Wilford Hall* is the name of the base hospital at Lackland AFB in San Antonio.) After his father recovered, Ken Compton kept prodding and asking his father about the statement, wanting to know more. Late in life, his father finally relented and told him that in 1964 he had seen two alien bodies on examination tables at Wilford Hall. They were small, 3 to 4 feet tall, and grayish in color. He did not tell his son how he came to view the bodies other than to say it was in an official Air Force capacity as an intelligence officer and that he had been ordered to attend the viewing by his superior. He also did not elaborate on anything beyond his brief description of their appearance. Ken said that his father was in the Air Force from 1952 to 1973 and, being from San Angelo, Texas, he also retired from the Air Force in San Angelo. Ken is also sure that he saw Sheridan Cavitt at his house on occasion.[25] Cavitt was also from San Angelo. An interesting sideline to this story is that seven years before we interviewed Ken Compton, we interviewed the widow of a man who worked at Randolph AFB in San Antonio, Texas. According to her, her husband came home from work one day in a somber mood. She asked him if he was all right, and he responded that he was just upset because he had seen something at work that he wasn't supposed to see. He said that he had been walking down the hallway in the aero medical facility at Randolph in the normal course of his work when he saw two little "beings from another world" being wheeled on gurneys into an examination room. He believed that they were dead from their grayish coloring, but it was their oversized heads for their diminutive stature and strange, wide-set eyes that bothered him. The year was 1964.[26]

Allow us to pull rank for a moment and add one final example to this growing list of most highly relevant witnesses. Sadly, time ran out for him, his family, and for us. And so we challenge even the scoffer to imagine just what information the next gentleman may have taken with him. As for pulling rank, this individual was awarded the French Croix de Guerre in 1943, the Legion of Merit and the Distinguished Flying Cross with One Oak Leaf Cluster in 1944, the French Croix de Guerre with Palm in 1945, the Order of the Crown of Belgium with the rank of Commander and the Air Medal with Four Oak Leaf Clusters in 1947, and, most prestigiously, the Order of Orange-Nassau with Swords of Netherlands, the Aldon Heron Medal of

Merit of Ecuador, the Presidential Medal of Merit of Nicaragua, the Geraldo O'Higgins Medal of Merit of Chile, and the U.S. Air Force Distinguished Service Medal in 1948. In 1946, he became a member of the Joint Operations Review Board of the Army and Navy Staff College and also joined the faculty of the National War College in Washington, D.C., as an instructor for strategic air operations; in 1947, he was assigned to Army Air Forces Headquarters in Washington, D.C., as the executive officer to the chief of staff, General Carl Spaatz; from 1948 to 1953, he served as the first U.S. Air Force aide to the president. In this capacity, he represented the Air Force in the White House and served as an informal consultant to the president on Air Force matters, which included quarterly briefings on *flying saucers*. After his White House tour of duty, he would go on to become the commanding officer at three Air Force bases, as well as returning to Washington as the Deputy Chief of Staff for Personnel; he retired in 1962 with thirty years of service that also included being a Rated Command Pilot with 7,500 hours flying time and flying combat missions in P-47 fighters and B-17 bombers in the European Theater during World War II.[27] His name: USAF Major General Robert Broussard Landry. His connection to Roswell: General Landry was Air Force aide to President Harry Truman at the time of the incident, and Landry would brief the president *verbally* on the "*flying saucer* situation," not leaving any paper trail.[28] Before he would die, he promised his family that he would "... tell them the whole story . . . the truth of what really happened at Roswell." But, as stated earlier, death is the great silencer, and the general lost his battle with time, with the result that we are all the less knowledgeable for that fact. Still, one ominous remark should haunt us, as recounted by his grandson Dave, when he was asked by his sister, Dr. Meg Blackburn Losey, "What else did Granddad tell you?" "Granddad told me, 'Dave, if I ever told you the truth of what happened back at Roswell, you would never see life in the same way again.'"[29] Do you think the general was *really* referring to balsa wood, balloons, and tinfoil? Mogul pales into a richly deserved insignificance.

Are we to believe that any of the aforementioned examples were the results of mere weather balloons and wooden crash dummies? It is with a great sense of bewilderment that we ask, why do all of these witnesses—to the supposed mundane—clearly have an obsession with "small bodies"? Should we be shocked that nary a one has ever suggested they were made of wood, as the military would want us to believe? Are we to conclude that they are all lying, deceiving their loved ones at the end of their lives? To the surviving families, it remains a feeble, futile exercise at best. And for those few whose true love of family would inevitably outweigh love of country—though it may have taken a lifetime of denial—we strongly maintain that their dying words meet all the criteria for reasonable doubt.

24

A VOICE FROM THE GRAVE:
THE SEALED STATEMENT OF FIRST
LIEUTENANT WALTER G. HAUT

I t is debatable whether a deathbed confession carries more weight than a signed testament. Legally, the latter is much more binding, in that it is a lawfully signed and witnessed document intended as a final statement. The individual is ensuring the transfer of personal possessions, or in the case of controversial or secret information, the creation of a sacred trust with particular beneficiaries to place in their custody private material to ensure that their final wishes are carried out. In most situations, immediate families serve as custodians, until after the testament author's death, of the property held in confidence to facilitate his or her last desires. In this chapter, we will reveal First Lieutenant Walter G. Haut's signed testament about his experience in Roswell in 1947.

During WWII, Haut was a B-29 bombardier on twenty bombing raids over Japan. Haut received numerous medals, including the Purple Heart, and in the summer of 1946, he was assigned with dropping the measuring instruments during Operation Crossroads' atomic tests at Bikini Atoll in the South Pacific. It is therefore fascinating that what Haut is most famous for was that, as the public relations officer for the 509th Bomb Group at the RAAF, he released to the media one of the most famous press releases of all time. And for the next fifty-five years, that's about *all* he would admit about his involvement. Finally, in November of 2002, Walter Haut declared his intentions regarding specific information he possessed about the Roswell Incident of 1947.

To say that the revelations were illuminating to his surviving family would be an understatement. Truly, they were shocked. In an atmosphere of ridicule and rejection of Roswell testimony from other witnesses, Walter Haut had chosen instead to eternally preserve his story as a solemn vow. But it was a vow that involved someone outside his immediate family.

We had the pleasure of meeting Walter Haut the very last day of our initial visit to Roswell, in February 1989. This was our anticipated one trip to New Mexico, where we fully expected to solve the entire mystery in a single investigative jaunt. We confidently set out to confirm the weather balloon explanation—or something just as earthly. But after having met with Mack Brazel's son Bill and his wife, Shirley, then with Walter Haut and his wife, "Pete," we seriously started to consider that we may have been wrong. Little did we realize at the time just how wrong.

Haut's story was at that time, as it remained for the next fifteen years, that he received orders from his boss, Colonel Blanchard, to put out a press release. He typed it up and distributed it to the local media. End of his involvement. However, what shined through all of our remaining questions about the

First Lieutenant Walter G. Haut was the RAAF's P.I.O. (Public Information Officer) in 1947. (Photo courtesy of the Army Air Forces.)

RAAF's supposed blunder over making such an outlandish claim was the deep and loyal friendship Haut had with the colonel up until his death. In fact, Haut fondly spoke of him as though he were a "close uncle." A look of admiration and respect would come over his face whenever he referred to Colonel Blanchard—as he affectionately called him, "the old man."

Certainly, it was not at all uncommon for a high-ranking military man such as Blanchard to take a young officer under his wing and lead him up the ladder. But who was "Butch" Blanchard to Walter Haut? What place does he have in American history? And was he as incompetent as Roswell skeptics would want us to believe? As one of the Air Force debunkers said, attempting to denigrate Blanchard's character and reputation, "Blanchard was a loose cannon!"[1]

Blanchard graduated from West Point in 1938. From that moment on he was placed on a fast track for promotion and by the end of WWII was perceived as a role model for the future of the Army Air Forces. As the deputy commander of the 58th Bomb Wing, he flew the very first B-29 into China to establish strategic bombing operations in the Japanese islands in 1944. As preparations for a full-scale invasion of Japan commenced, Blanchard was next assigned as commander of the 40th Bomb Group, B-29 squadron, and subsequently as the operations officer of the 21st Bomber Command. This is where he and General Curtis LeMay prepared and supervised plans for the first atomic bomb to be dropped on Hiroshima. With orders for a second bomb, Blanchard was originally assigned to pilot the flight over Nagasaki. Colonel Charles Sweeny would command the B-29 *BocksCar* in his place. With the surrender of Japan, he was General Roger Ramey's right-hand man in 1946 as the CO of the 509th Atomic Bomb Group in Operation Crossroads. One of the officers he oversaw was Walter Haut, with whom he quickly became good friends.

After his tenure at Roswell, Blanchard was assigned to the Strategic Air Command, Eighth Air Force Headquarters, as director of operations in 1948. From there he was in charge of the atomic training of B-36 crews, our first intercontinental bombers. After commanding B-50 and B-36 bomber units of SAC, he was made deputy director of operations for that unit in 1953. Next, he would become inspector general for the Air Force, then deputy chief of staff for programs and requirements at AF headquarters in Washington, which led him to AF planning and operations. He would also become a senior AF member of the Military Staff Committee of the United Nations. And lastly, he was vice chief of staff of the USAF and certain for the Joint Chiefs. By that time he had reached four stars, but sadly died at his desk at the Pentagon in 1966 from a heart attack. General William Blanchard, who couldn't identify a common weather balloon at Roswell in 1947, achieved all of this and more by the age of fifty!

Is it any wonder that Haut had the utmost esteem for and confidence in his military boss as the ultimate role model? In fact, there is little doubt that Haut could have followed him all the way to the rank of general. Such was the personal bond between these two men. When "the old man" needed someone to drive his wife and kids back home to Chicago, Haut was his man. When the base commander was touring some Washington dignitary around the base, Haut was always at his side (while Ramey followed from behind). And when Blanchard suddenly died, Haut didn't read about it in the paper, and he didn't get a phone call from some military underling. A special emissary arrived in Roswell from Washington and went to Haut's home with the somber news. This was normally a courtesy afforded only immediate family. And to Blanchard, that is exactly what Haut was.

We have explained the close association of these two men in order to address this continuing argument: Walter Haut basically kept silent for more than fifty years; why profess any hidden truths for posthumous disclosure? Why withhold any details about such an important story? Why not write a book? Why not at least profit your family? After all, isn't it human nature for one to have milked this account for all it was worth? The more simple answer as to why he never spoke to a fuller extent about Roswell was because *he promised Butch that he would not.* To Haut, it was a matter of his word to the most trusted friend and commander he had in his life. It was his life-long testimonial as well as an indication of just how highly he regarded his former commanding officer. True, national security, protection of loved ones, and secrecy oaths may have played into his decision. But, more importantly, Blanchard had asked him not to say anything.

The reader should be reminded that Blanchard himself took the entire 1947 ordeal very seriously. His first wife, Ethel, recalled that he had come home in a very agitated state and told her that a flying saucer had crashed in the desert near Roswell. He told her about the little bodies: "They're human, but they're not," and the strange writing, "...looks like Russian, but it isn't." She also added that her husband had been "very upset and beside himself."[2] According to Blanchard's daughter, Dale, her father's demeanor was normally just the opposite. "My father wasn't a person who got upset. He was always calm, and seeing my dad agitated was rare. He had a plodder's personality. He did not speak quickly, and weighed his words."[3] As described here, the similarity of Blanchard's personality to that of Walter Haut's is remarkable.

There were always subtle hints in Haut's life that he was withholding secret information or protecting someone in particular. For example, he would receive many Christmas cards from the former head of CIC at Fort Worth, Milton

Knight. After World War II they came home from Saipan on the same airplane. Knight would follow in Ramey's footsteps by becoming part of Project Blue Book. A rather curious relationship developed, with Knight paying Walter visits at his home in Roswell at least half a dozen times from 1952 through 1956. Each time, Knight would arrogantly display pictures of UFOs that were, according to Haut, "always explainable." We will leave this to the psychologists

Lieutenant Walter Haut and Colonel Blanchard with Secretary of War Robert P. Patterson on his inspection trip to the RAAF in early 1947. SAC Commanding General, George Kenny, and SAC Deputy Commander, Clements McMullen (partial head only) can be seen walking behind. (Photo courtesy of the 1947 RAAF Yearbook.)

to diagnose, but to Walter it was quite obvious that Knight's intent was to create doubt in the mind of the former RAAF PIO. Add to that the odd Christmas card that we were shown by Walter in 1989. It included a note that read, "I still say that there were no bodies at Fort Worth." The inference was quite obvious. Unfortunately, Walter couldn't elaborate any further as to what had set off the debate. From our first meeting it was clear that all of his experience in dealing with the press made him quite proficient at playing dodgeball.

Instead, Mr. Haut preferred to play it safe by remaining on the periphery. He was always happy to pass on limited information to us about others who were involved, maybe to temporarily avoid renewed focus on himself. Our position has always remained simple: Colonel Blanchard, who was the base commander and who would have been privy to every aspect of the case + Walter Haut, his most trusted ally and friend = Walter Haut must possess a great deal more information than he's admitting to us.[4] But he had made a promise. Albeit reluctantly, we couldn't help but admire the man's loyalty.

Ironically, it was inevitable that Walter Haut would become the greatest champion of all the believers in Roswell. His testimony was always consistent as to the limited capacity of his participation. But at the end of each public appearance or media interview, he would always hit it out of the park with his closing remarks: "It wasn't any type of weather balloon," he would say. "I believe it was a UFO! Just don't ask me why."[5] But being the intrepid investigators we are, we persisted. After all, Haut had given us the final solution to the longstanding mystery—he told us that he believed it was a UFO. How did he come to accept that? What did he really know?

Keeping all of this in mind, we would be remiss if we did not mention a much more ominous reason for Haut's reluctance to speak out for all those years. It fits the pattern we have heard for both civilian and military alike as to the cause of witness silence. Walter had been receiving regular phone threats for many years after the incident. "There were so many calls, I lost track of them—*about thirty years of it!*" a disgusted Haut would complain. Long before caller ID, he got menacing calls from the "Norseman" in Canada who would shout "HOWT!" into the phone and then demand that he forget the whole matter. "I was told to forget it—or else!" Another, more direct caller who identified himself as "The Shadow" would harass the Hauts at all hours of the night. "Each and every time he called, he told me that my life was in jeopardy if I talked about the incident," said Haut. In the 1990s, he also started receiving letters from the son of the late General William C. Kingsbury. In 1947, Lieutenant Colonel Kingsbury was the commanding officer of the 715th Bomb Squadron at the RAAF. He would later become the deputy commanding officer of the 509th Bomb Wing at Roswell and then go on to a host of other assignments in his career, retiring in 1966 as an Air Force major general. He passed away in 1973. Although Walter Haut was in his seventies by the 1990s, Kingsbury's son, a retired Air Force colonel, addressed Haut as a subordinate in his letters to him (in other words, "lieutenant"), while delivering the threat that "lieutenants should know to keep their mouths shut." Is it any wonder that there remained circumstances in Haut's life that weighed heavily on whether to break his silence or not?

Of our close to one hundred research trips to New Mexico before Haut died, he was almost always included in our rushed schedule. But beyond our fading hope that he would someday take us into his confidence, mainly, he was our friend, and he was Roswell's greatest ambassador. Not

Walter Haut in a mid-1990s photo. (Photo courtesy of Tom Carey.)

that he ever claimed the Roswell Incident really happened. In many regards he was standing up for all his military comrades long passed and forever silent. He also remained one of our last connections to one of the most important events in history. We just had to come up with a respectable device that would enable him to preserve the truth—without breaking his word to his departed friend.

When Haut founded the world-famous International UFO Museum & Research Center in his chosen home of Roswell, we believe that somehow, deep down, he was doing what the old man would have wanted him to do. He was still keeping his word; he was not breaking any promise. He provided a public repository for information related to the event, and, more importantly, he established a facility for the preservation of the Roswell Incident as a historic event. Such may have been the reasoning for others, such as Major Patrick Saunders just before he passed away—let others discover the facts, and when the time is appropriate you write in their book, "Here's the truth and I still haven't told anybody anything." (See Appendix III.) They are free to leave us with a clear conscience.

We would like to truly believe that this was the opportunity we presented to Haut. By preparing a sealed statement to be released after his death, his true legacy pertaining to Roswell is now as complete as it can ever be. The verisimilitude known as the Roswell Incident has become all the clearer thanks to Walter Haut. And through his emotional evolution we would like to believe that he could look his old friend in the eye and say, "As long as I lived, I kept my word."

But, as we should all profess, the truth is the truth, and the facts are the facts. And no government, no organization, and no one individual, whether advocate or debunker, has been given jurisdiction over defining what we are to accept as reality—a concept that would allow for those rare occasions when secrets do come from the grave.

Thank you for all the Christmas cards, Walter.

The following is the sealed affidavit of First Lieutenant Walter G. Haut. It is presented here for the reader's consideration with the permission of his surviving family, unedited and without notes or editorial comment. It is provided for posthumous release for the sole purpose of being placed in the historic record by Mr. Haut, as was his final request with the sincerest of motives. Any assertions to the contrary are without merit. No reproduction of this statement can be granted without written request for permission from his surviving family.

SEALED AFFIDAVIT OF
WALTER G. HAUT
DATE: December 26, 2002 WITNESS: Chris Xxxxx
NOTARY: Beverlee Morgan

1. My name is Walter G. Haut.

2. I was born on June 2, 1922.

3. My address is 1405 W. 7th Street, Roswell, NM 88203

4. I am retired.

5. In July, 1947, I was stationed at the Roswell Army Air Base in Roswell, New Mexico, serving as the base Public Information Officer. I had spent the 4th of July weekend (Saturday, the 5th, and Sunday, the 6th) at my private residence about 10 miles north of the base, which was located south of town.

6. I was aware that someone had reported the remains of a downed vehicle by midmorning after my return to duty at the base on Monday, July 7. I was aware that Major Jesse A. Marcel, head of intelligence, was sent by the base commander, Col. William Blanchard, to investigate.

7. By late in the afternoon that same day, I would learn that additional civilian reports came in regarding a second site just north of Roswell. I would spend the better part of the day attending to my regular duties hearing little if anything more.

8. On Tuesday morning, July 8, I would attend the regularly scheduled staff meeting at 7:30 a.m. Besides Blanchard, Marcel; CIC Capt. Sheridan Cavitt; Col. James I. Hopkins, the operations officer; Major Patrick Saunders, the base adjutant; Major Isadore Brown, the personnel officer; Lt. Col. Ulysses S. Nero, the supply officer; and from Carswell AAF in Fort Worth, Texas, Blanchard's boss, Brig. Gen. Roger Ramey and his chief of staff, Col. Thomas J. DuBose were also in attendance. The main topic of discussion was reported by Marcel and Cavitt regarding an extensive debris field in Lincoln County approx. 75 miles NW of Roswell. A preliminary briefing was provided by Blanchard about the second site approx. 40 miles north of town. Samples of wreckage were passed around the table. It was unlike any material I had or have ever seen in my life. Pieces, which resembled metal foil, paper thin yet extremely strong, and pieces with unusual markings along their length were handled from man to man, each voicing their opinion. No one was able to identify the crash debris.

9. One of the main concerns discussed at the meeting was whether we should go public or not with the discovery. Gen. Ramey proposed a plan, which I believe originated with his bosses at the Pentagon. Attention needed to be diverted from the more important site north of town by acknowledging the other location. Too many civilians were already involved and the press already was informed. I was not completely informed how this would be accomplished.

10. At approximately 9:30 a.m. Col. Blanchard phoned my office and dictated the press release of having in our possession a flying disc, coming from a ranch northwest of Roswell, and Marcel flying the material to higher headquarters. I was to deliver the news release to radio stations KGFL and KSWS, and newspapers the *Daily Record* and the *Morning Dispatch*.

11. By the time the news had hit the wire services, my office was inundated with phone calls from around the world. Messages stacked up on my desk, and rather than deal with the media concern, Col. Blanchard suggested that I go home and "hide out."

12. Before leaving the base, Col. Blanchard took me personally to Building 84, a B-29 hangar located on the east side of the tarmac. Upon first approaching the building, I observed that it was under heavy guard both outside and inside. Once inside, I was permitted from a safe distance to first observe the object just recovered north of town. It was approx. 12 to 15 feet in length, not quite as wide, about 6 feet high, and more of an egg shape. Lighting was poor, but its surface did appear metallic. No windows, portholes, wings, tail section, or landing gear were visible.

13. Also from a distance, I was able to see a couple of bodies under a canvas tarpaulin. Only the heads extended beyond the covering, and I was not able to make out any features. The heads did appear larger than normal and the contour of the canvas over the bodies suggested the size of a 10-year-old child. At a later date in Blanchard's office, he would extend his arm about 4 feet above the floor to indicate the height.

14. I was informed of a temporary morgue set up to accommodate the recovered bodies.

15. I was informed that the wreckage was not "hot" (radioactive).

16. Upon his return from Fort Worth, Major Marcel described to me taking pieces of the wreckage to Gen. Ramey's office and after returning from a map room, finding the remains of a weather balloon and radar kite

substituted while he was out of the room. Marcel was very upset over this situation. We would not discuss it again.

17. I would be allowed to make at least one visit to one of the recovery sites during the military cleanup. I would return to the base with some of the wreckage which I would display in my office.

18. I was aware two separate teams would return to each site months later for periodic searches for any remaining evidence.

19. I am convinced that what I personally observed was some type of craft and its crew from outer space.

20. I have not been paid nor given anything of value to make this statement, and it is the truth to the best of my recollection.

THIS STATEMENT IS TO REMAIN SEALED AND SECURED UNTIL THE TIME OF MY DEATH, AT WHICH TIME MY SURVIVING FAMILY WILL DETERMINE ITS DISPOSITION.

Signed: Walter G. Haut
Signature Witnessed by: Chris Xxxxxx
Dated: December 26, 2002

POSTSCRIPT

As a final postscript, the U.S. Department of Justice has determined that voice analysis testing such as a Psychological Stress Evaluation (PSE) test can be up to 97 percent accurate. Surpassing a PSE, the new state-of-the-art development is called Truth Verification and Credibility Assessment utilizing Artificial Intelligence (AI) through Digital Voice Stress Analysis (DVSA). Using space-age technology the AI determines voice patterns down to the slightest nuance and assigns code numbers for the computer to evaluate through algorithms the validity of the sample provided. A video interview of Haut, recorded in 2000, was provided to Baker Group International for such elaborate testing. There, the company founder and CEO E. Gary Baker PhD., considered the world expert in this field, analyzed the testimony of Walter Haut in early 2021. Dr. Baker concluded that Haut's descriptions and eyewitness testimony of the Roswell wreckage, craft, and bodies were truthful. Modern technology has further substantiated the credibility of Walter Haut.

25

SEARCHING FOR ROSWELL'S
HOLY GRAIL

No aspect of the Roswell investigation has been more frustrating than the search for physical evidence of the crash. Even though there is no doubt in our minds that we would win the case, legally, in a court of law based solely upon the weight of our witness testimonies vís-a-vís the Air Force's Project Mogul and anthropomorphic dummies-from-the-sky explanations for Roswell, we could not heretofore prove the case, scientifically, due to a lack of irrefutable physical evidence. It has been the one shortfall in our case to date that has sustained the hardcore skeptics in clinging to their belief system in the face of overwhelming evidence for the reality of Roswell. But that may change, and soon.

We have likened our search for physical evidence from the 1947 Roswell UFO crash to that of the Holy Grail—the Medieval legend pertaining to the cup used by Jesus at the Last Supper, for which biblical researchers and archaeologists are still searching today. Way back in the Introduction to this book, we identified our search for the "Holy Grail of Roswell" as a search for a piece of "memory metal" that has been described by so many of the witnesses who had the good fortune to handle a piece of it. Throughout the years, we have logged thousands of miles in this search, based as they were on what turned out to be false leads. In one case, a Roswell plumber had told a relative at a family get-together of coming across an old Army footlocker inside a cement wall in the basement of a home that he had to knock down to run pipe. He told his relative that inside the footlocker he found old Army items such as a .45 caliber pistol, an Army officer's hat, and *something else*— very thin, palm-sized pieces of metal that he could wad up in one hand and, when he opened his fist, would instantly resume their original shape. When the relative, who was a friend of ours, called us with the story, we dropped everything and were on the next flights to Roswell. Somehow, we talked the lady who owned the house in question into taking us down into her dank basement. The excitement that had been building as we descended each step came to screeching halt when she told us, "Yes, I remember the plumber, but he never went into the basement. He only fixed my kitchen sink." The truth of her statement was then borne out when we examined the basement walls, which showed no signs of any recent activity. All attempts by our friend to reach his plumber relative after that proved futile as phone calls and messages went unanswered. A joke? A mistake? We can only guess.

We have been given leads from any number of people "who know someone who knows someone who..." that have all come up short of reaching that *someone* who allegedly has the goods. Then, when we attempt to reestablish contact with our original sources, they disappear by not answering the phone. We know that routine to a T. About ten years ago, the UFO Museum in Roswell offered a $1 million reward for a piece of physical evidence that could be proven to have been manufactured off the planet. Much was promised, but not much was delivered to the UFO Museum. The most notable submission was a strange-looking piece of metal that, upon analysis, turned out to be a piece of Japanese jewelry! It had originated with a jeweler in Utah and then passed through several intermediaries before showing up in the UFO Museum's in-basket. A joke? A mistake? We've been there. The wayward piece of jewelry is now hanging on the wall at the UFO Museum as a curiosity item. We have also been told that there are elderly Roswell or Corona ranchers out there who, upon their passing, have promised to give up their pieces of crash wreckage they have been secretly keeping since 1947. Although the ranchers' names are never given, these stories do have some element of truth to them, as we believe that there *are* pieces of wreckage from the 1947 crash still out there in the possession of ranchers who got to the crash sites before the military arrived. Unfortunately, they keep passing, but nothing has shown up. A joke? Mistake? Or someone just trying to impress by speaking out of turn? Been there too.

Our most recent experience along these lines involved the son of an RAAF corporal who we know for a fact was heavily involved in the crash/recovery. He had been cited by name by other eyewitnesses who had been there, and his picture is in the 1947 RAAF yearbook. According to his son, who had contacted us in March 2008, his late father had pocketed some of the debris that included some small "crystals" as well as "memory metal" during the cleanup, which he kept in a small, wooden box in the basement of his home. He said that his father was also a photographer who had taken pictures—lots of them—at the crash site, of the craft, and of the bodies. He even had his father's old camera, and there was still dirt in the box! We definitely were on the next plane—almost! The son stopped us by saying that he first had to visit his father's brother in California to retrieve additional items that were his father's. When asked why he was coming forward with this information and artifacts now, he replied, "It's time. I would have come out earlier, but I was given advice by family members not to." The time frame for his trip to California came and went. When he didn't call us, we tried to reach him. All efforts to reach him since have been met with a stony silence. A joke? Mistake? Or simply a case of cold feet? You make the call.

Other researchers throughout the years have presented bits of metal alleged to have come from the Roswell crash. These usually disappear into a black hole of endless testing by "scientists" with questionable credentials, and are never heard about again.

In 2002, with funding from the Syfy Channel under supervision of the University of New Mexico's Office of Contract Archaeology, we were able to employ a more scientific approach to our quest for physical evidence by conducting a full-scale archaeological dig at the Mack Brazel/J. B. Foster Ranch debris field site near Corona in Lincoln County, New Mexico. Everything was conducted in accordance with standard archaeological field practices (laying out a grid, digging test pits, cataloging, photographing, and securing each recovered item, and so on). Unfortunately, less than half of the promised dig volunteers showed up, and as a result, less than 10 percent of the total site area was actually excavated. Still, a number of anomalous "artifacts" were unearthed in the vicinity of the alleged gouge (a 500-foot-long scar in the terrain said to have been cut by the UFO as it "skipped" before becoming airborne again).

Setting up for the 2002 archaeological excavation of the Brazel debris field site on the old J. B. Foster Ranch that was funded by the Syfy Channel. (Photo courtesy of Tom Carey.)

The professional archaeologists could not identify these, but assumed that they must be something "natural" rather than extraterrestrial.[1] A cut across the alleged trajectory of the gouge with a backhoe (like cutting a layer cake) also yielded a surprising "artifact," evidence to us of a V-shaped gouge that had been cut into the surface of the Earth now filled by years of wind-blown and water-borne topsoil, right where witnesses said it would be! The archaeologists remain uncertain, however, preferring to identify it as a "superficial anomaly."[2]

A followup dig in 2006 for the Syfy Channel's TV show *Sci Fi Investigates*, with even fewer diggers than before, digging in the same place as before, found nothing of significance. The highlight was during an unexpected demonstration of how far from the immediate site area some of the lightweight crash debris may still be lying in wait to be discovered, when all of a sudden a whirling dust devil swooped down like a mini tornado, seemingly out of nowhere, scooped up the producer's papers, and carried them in its swirling plume hundreds of feet into the air, finally depositing them miles away. We could do nothing but watch in awe and think back to 1947.

Life can surprise you, though. Sometimes, when one door closes on you, another door unexpectedly may be opening. And so it has been with our Roswell investigation in the person of Anthony Bragalia. A self-employed businessman,

Bragalia has been conducting his own private investigation of the Roswell crash/retrieval for a number of years. We feel privileged that he has chosen this book to publish the fruits of his many years' research into one phase of the Roswell story that has received scant attention from researchers: What happened to the crash wreckage after it was delivered to Wright AAF in Dayton, Ohio? His research into this area of inquiry has been outstanding, and the results achieved so far have been astounding, in the form of **scientific, documentary evidence proving the extraterrestrial nature of the Roswell wreckage!** The newly discovered documents and their references

The leaders of the 2002 "dig" were [L to R] coauthor, Don Schmitt, UNM archaeologist Dr. William Doleman, the Sci-Fi Channel's Larry Landsman, and coauthor Tom Carey. (Photo courtesy of Tom Carey.)

to highly classified government reports on "memory metal" show that the Air Force—through Battelle and other laboratories, such as at New York University (of Project Mogul fame)—was attempting to replicate the "shape recovery" properties of the Roswell crash debris!

It started with a former Battelle employee who confessed to a family friend in 1960 that he had been tasked to analyze crashed flying saucer debris when he worked at Battelle. We had always suspected that the Battelle Memorial Institute of Columbus, Ohio, because of what it does and where it's located (close to Wright-Patterson AFB where the Roswell wreckage was taken), would have been called upon by the Air Force to conduct studies on the wreckage. But we never did anything beyond speculating about it. Bragalia has done something, and he has proved our earlier speculations to have been on target by uncovering actual documentation confirming Battelle's analysis of the Roswell debris! According to Bragalia, he has found references in the scientific literature that cite a Battelle "progress report" that it issued to Wright-Patterson AFB that shows that, in the two years immediately following the Roswell crash in 1947,

Battelle was testing and developing material that was distinctly Roswellian in nature! We are talking "memory metal" here. Bragalia has been able to identify the authors of the two Battelle progress reports, written in 1948 and 1949, both of which remain highly classified within the inner sanctums of the Air Force at Wright-Patterson to this very day. The authors are now deceased, and to make matters even more difficult, all current and former Battelle employees remain under lifelong security oaths if they have worked on national security matters.

The results of the Battelle studies of the Roswell wreckage bore fruit in the form of our own "memory metal," known today as nitinol, an amalgam of nickel and pure titanium that displays "shape-recovery" properties similar to those described by Roswell witnesses. Nitinol "remembers" its original shape by returning to that shape when crumpled. It possesses a high fatigue strength, is lightweight, is aluminum-like in color, and can withstand a blowtorch. Apparently covering its tracks, the Air Force let the U.S. Naval Ordnance Lab "officially" introduce nitinol to the scientific community in the early 1960s. Today, nitinol is used in medical devices, antennae, bendable eyeglass frames, and many other applications. A variant of it is being developed by NASA to create a "self-healing" outer skin material for spacecraft. From Bragalia's report, "It is likely that nitinol is not itself the Roswell debris. Rather, it is more likely that it is the result of attempts by Wright-Patterson, through Battelle, to simulate the properties of the Roswell debris and exploit the obvious defense and technology value of 'morphing' or 'living' metals.[3] This is why UFOs are seen to 'morph'! ET has mastered a 'special process,' as General Exon alluded to. Their craft is composed of manipulated material that has remarkable shape recovery, memory, and other properties—properties that the Air Force tried to duplicate through Battelle!"[4] It also could explain why so much of the Roswell wreckage debris, as described by the witnesses, was composed of "memory metal."

Since the second edition of this book (2009), Anthony Bragalia has continued his intrepid quest to find out what became of the Roswell wreckage. Appendix IV details Bragalia's most recent research and what is perhaps his most exciting result to date—an acknowledgement by the U.S. Department of Defense that such wreckage exists, where it is located, and the nature of the experiments being conducted on it.

The prosecution rests—for now.

POSTSCRIPT
BY THOMAS J. CAREY

Not so long ago, before personal computers, iPads, cable TV, and social media, most people received their news of the day from radio, network television (ABC/NBC/CBS) and the print media (i.e., newspapers and weekly/bi-weekly news magazines like *Time*, *Newsweek*, and *U.S. News & World Report*). This was also a time prior to the development of sophisticated photographic software that could reproduce photographs sharply and clearly into the print media, especially newspapers, for publication. If you took a magnifying glass to any photo that appeared in your favorite newspaper of that time, you immediately noticed that it was not of the seamless quality that you were used to seeing in the photos you took from your own camera. The newspaper photo actually consisted of small dots of varying shades that, when viewed from just a few inches away, formed a recognizable photo image. Each dot, by itself, conveyed no information and meant nothing; but, connected to all of the other dots in the matrix, they "painted" an incontrovertible, recognizable picture.

Taking the idea of "dots-forming-pictures" to another level provided the basis for a popular quiz show of the late 1950s called *Dotto*. There were two contestants, each presented with an incomplete portrait of a famous person that consisted of a few filled-in areas of the face and head plus a number of empty areas that made the identity of the person in the portrait unrecognizable. The empty spaces were connected to the filled-in spots by a series of linear dots that, if connected, would render the portrait recognizable. Each contestant would be asked questions. For every correct answer, the contestant who answered it would get to connect a number of dots of his or her choice. This procedure would continue until one of the contestants thought they recognized the person in the portrait and hit the buzzer to give an answer. If the answer was correct, the game was over, and the losing contestant went home. If not, the game would continue until enough dots had been connected to convince one of the contestants that it was John Wayne and not Wayne Newton.

When we started our Roswell investigation in 1989, there were a lot of empty spaces that needed filling in and not enough solid areas to anchor around. We thought that the Plains of San Agustin near Socorro and Magdalena, New Mexico, 150 miles west of Mack Brazel's sheep ranch on the old J. B. Foster Ranch, was where the UFO had crashed. We thought that that's where the little bodies that had been reported were located as well. We thought that the archaeological team that had allegedly discovered the crashed UFO on "the flats," as the Plains of San Agustin was called by locals, was from the University of Pennsylvania in Philadelphia, Pennsylvania, under the leadership of an archaeologist by the name of "Dr. Adrian van Buskirk." We thought that the date of the UFO crash was possibly July 3, or possibly July 4, or maybe July 5 in the summer of 1947. None of these things turned out to be true.

Following the dictates of the German-American anthropologist Franz Boas's empiricism to "record enough facts, and the answer will fall to you like a ripe fruit," we have accumulated over the course of just over thirty years, one by one, approximately 600 first- and secondhand witnesses to form a framework of what took place near the town of Roswell, New Mexico, during the first week of July 1947. Applying a dose of "Occam's Razor" to Boas's "facts" and our witnesses' connected "dots" yields the most parsimonious conclusion to explain them as the one that General Ramey once described to someone as being, "out of this world!"

Over the years, a number of lawyers who have read our books have contacted us to tell us that if we hypothetically took our case for Roswell to court against, say, the U.S. Air Force or the Department of Defense—if that were possible—we would "slam dunk" win the case "hands down," take your pick. And further, we would win the case, not only by a "preponderance of the evidence" presented, as in civil cases, but also "beyond a reasonable doubt," which is the threshold in criminal cases!

We have often been asked, "Where are *they* from?" and, "Why are *they* here?" We can speculate regarding the second question that, since *their* arrival here en masse in June of 1947, perhaps it was our acquisition of nuclear weapons near the end of World War II that might have spurred an interest in us by *them*, but that is sheer speculation. The first question is the more interesting question of the two, but, given our mandate to find every last witness before Mr. Undertaker found him or her, we have never had that discussion. Thinking about it now, however, given the vast interstellar distances that must be traveled to get *here* from *there*, our visitors must have somehow overcome the "space-time continuum," by becoming, in essence, "time travelers." Such a conclusion has been offered by two acclaimed scientists who were heavily involved in the

Dr. Wernher von Braun, the "Father of America's Space Program," twice spoke publicly about the reality of UFOs. (Photo courtesy of the U.S. Army.)

Roswell case, Dr. Ernst Steinhoff and Navy Commander George Hoover.

Ernst Steinhoff was the second highest-ranking "Paperclip" German rocket scientist (after Wernher von Braun) brought over to the United States after World War II to help get our rocket program "off the ground." Along with von Braun, he was brought into the Roswell investigation and ultimately concluded that it was not a case of "extraterrestrial" visitation, but one of *extra-dimensional* visitation by "time travelers" able to "bend" or "warp" the space-time continuum to allow them to traverse the daunting distances of the universe.[1]

The brilliant George Hoover is known today as the "Grandfather of Satellite Technology." He was the Navy's first "UFO Guy" and a close associate of Wernher von Braun who was also heavily involved in the Roswell crash investigation. His conclusions regarding it are startling.

- ▶ Roswell was a crash event involving "visitors from *someplace else*."

- ▶ The entities were "not so much interplanetary as they were *extra-temporal* 'time travelers.'"

- ▶ The visitors are "from the future and may even be *us* from a future Earth."

- ▶ The visitors remain, however, "in essence *corporeal* and 'physical' to us."

251

There are a trillion galaxies in the universe, each with billions of stars with planets orbiting them, millions of them earthlike. (Photo courtesy of Wikimedia Commons.)

► Secret attempts at reverse-engineering the visitors' crash materials were made.

► The government feared the intentions and abilities of the visitors.[2]

Commander Hoover died in 1988, and Dr. Steinhoff died in 1979. Given their credentials, what they have told us about Roswell must give one pause, as must what their compatriot, Dr. Wernher von Braun said twice publicly about the *visitors*, "We find ourselves faced by powers, which are far stronger than hitherto assumed, and whose base is at present unknown to us."[3]

As our friend and colleague Stanton Friedman opined early and often in the investigation of the Roswell case, "We are racing the undertaker [to interview as many of the witnesses as we can before 'Father Time' claims them]." Sadly, Stan, too, is now gone. His motto has always been our motto—to locate and interview as many firsthand witnesses as we could before they were all gone. As this is written, we can think of only one firsthand witness who still *may* be alive, and last we heard, he was in a Roswell nursing home with Alzheimer's. There still are a number of things on which we want to follow up, but the *proactive*, "gumshoe," witness-chasing phase of our Roswell investigation has come to an end. The undertaker may have won a major battle, but we will still win the war.

AFTERWORD
BY GEORGE NOORY

A t the time of this writing, it is July 2009, sixty-two years after the crash outside Roswell, New Mexico, of some kind of flying craft. Because it is sixty-two years after the event, many of the eyewitnesses, some of whom were alive at the fiftieth anniversary, have since died. For some, their stories died with them.

For men like Lieutenant Colonel Marion Magruder and Lieutenant Walter Haut, their stories live on in statements they made that constitute deathbed confessions. For some of the other witnesses, their children, grandchildren, or friends have added to the tale with what they said before they died.

These statements, particularly those collected by Thomas J. Carey and Donald R. Schmitt in the updated and revised edition of their bestselling book, *Witness to Roswell*, not only shed additional light on the case, but, in my opinion, blow away any arguments skeptics and debunkers can put up. When I finished their book, I said to myself, "Case closed!" for the very wealth and sheer weight of eyewitness testimony.

Carey and Schmitt have published some remarkable revelations that any open-minded, objective individual would have to take note of and not just casually dismiss: new firsthand accounts from Mac Magruder, Major Patrick Saunders, General Arthur Exon, and even from legendary Roswell witness Jesse Marcel. Their stories stand as a testament to the veracity of the Roswell Incident. But two years ago we learned of yet another witness, a witness we knew and spoke to, but we didn't know his real story until Don Schmitt and Tom Carey wrote it up in *Witness to Roswell*.

This, of course, is the real story of Lieutenant Walter Haut, who, at Don Schmitt's encouragement, had sworn out a sealed statement on December 26, 2002, telling the complete story of his involvement with the Roswell crash. The revelation of this statement and its contents were nothing short of a blockbuster, and, as I said publicly on my radio program, no one reading this statement will ever doubt the veracity of Roswell again.

Walter Haut's legacy is what he left us after he passed away. It is testamentary evidence to the truth of the Roswell cover-up as well. It describes a plan agreed to by officers at the Roswell Army Airfield, but perhaps dictated by the higher-ups on the Army general staff or even the White House. But reading Haut's last testament, written in plain, simple, and direct language, makes the hair stand up on the back of your neck. Sure, debunkers and skeptics can go after it. But in my opinion, there's nothing there to attack. Haut is not writing a piece of disinformation. He actually lived a life of disinformation for almost sixty years after Roswell, starting the Roswell Museum even as he knew the absolutely true story of what had happened, because he was part of it. Eyewitness testimony is the life's blood of the entire judicial system, and Roswell should be no exception.

The truth is clear and simple. A UFO crashed outside of Roswell, New Mexico, in 1947, and many of the witnesses exposed in this eye-opening book by Carey and Schmitt were participants in the cover-up of what actually happened; at the same time, they unwittingly set up the Army's plausible denial that there even was a UFO crash in the first place. The statement is a magnificent coda to the life of Walter Haut, coming as it does after his life was over. I think it tells the truth and dispels any doubt about what happened at Roswell. Case closed!

—GEORGE NOORY, host, *Coast to Coast AM*

Our investigation into the true nature of the events surrounding the apparent visitation of the planet Earth by extraterrestrial entities in 1947 continues apace. This volume has attempted to present to the reader not a complete overview of the Roswell case or even our own investigation of it, but snapshots of some of the seminal events as they happened at the time, on a more personal level to the individuals involved. We hope that readers will appreciate the approach we have taken in this volume. The facts presented herein are as they have been described to us or as determined by our decades-encompassing investigation. The interpretations accorded these facts are solely those of the authors.

The authors may be contacted through the offices of the International UFO Museum & Research Center in Roswell, New Mexico, located at:

114 North Main St.
Roswell, NM 88203
(800) 822-3545 or (575) 625-9495 [local]
(575) 625-1907 [fax]
deputydirector@roswellufomuseum.com

Or directly to the authors at:

(267) 722-8121
(267) 722-8121 [fax]
tcarey1947@aol.com
donald.schmitt@att.net
www.roswellinvestigator.com

APPENDIX I

MAP OF NEW MEXICO KEY POINTS

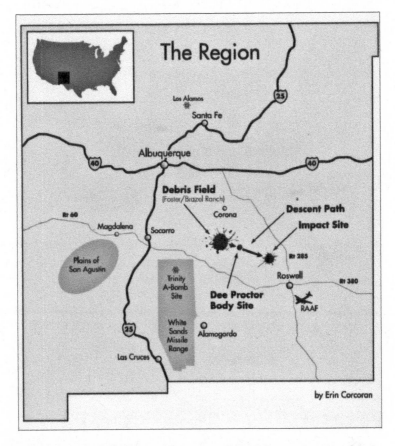

Schematic map of New Mexico showing the three 1947 Roswell-related crash locations in relation to key New Mexico cities, highways, and other landmarks. (Illustration courtesy of Tom Carey and Erin Carey Corcoran.)

APPENDIX II

CRASH SITE STONE MARKER

The historical stone marker on the J. B. Foster ranch was commissioned by the Syfy Channel and dedicated at the debris field [crash] site in 2005. (Photo courtesy of Tom Carey.)

The inscription reads:

IN JULY OF THE YEAR 1947 A CRAFT OF UNKNOWN ORIGIN
SPREAD DEBRIS OVER THIS SITE. WITNESSES WOULD
REPORT MATERIALS OF AN UNEARTHLY NATURE.

IN SEPTEMBER OF THE YEAR 2002 THE SCI FI CHANNEL
BROUGHT SCIENTISTS FROM THE UNIVERSITY OF NEW
MEXICO TO SEARCH THIS GROUND FOR EVIDENCE OF
THAT FATEFUL NIGHT.

BE IT OBSERVED THAT WHATEVER THE TRUE NATURE OF
WHAT HAS RESPECTFULLY BECOME KNOWN AS THE ROS-
WELL INCIDENT, HUMANKIND HAS FOREVER BEEN DRAWN
TO THE STARS.

DEDICATED JULY 5, 2005

APPENDIX III

MAJOR PATRICK SAUNDERS'S CONFESSION

Shortly before his death in 1995, the former RAAF base adjutant in 1947, Major Patrick H. Saunders, purchased a number of copies of the newly released paperback edition of *The Truth about the UFO Crash at Roswell* by Kevin Randle and Don Schmitt. Saunders then sent copies of the book around to close friends and family with a handwritten note on the flyleaf, as shown here. (Image of Saunders's note courtesy of Kevin Randle.)

Here's the truth and I still haven't told anybody anything!

DAMAGE CONTROL

Files were altered. So were personal records, along with assignments and various codings and code words. Changing serial numbers ensured that those searching later would not be able to locate those who were involved in the recovery. Individuals were brought into Roswell from Alamogordo, Albuquerque, and Los Alamos. The MPs were a special unit constructed of military police elements from Kirtland, Alamogordo, and Roswell. If the men didn't know one another, or were separated after the event, they would be unable to compare notes, and that would make the secret easier to keep.

After the impact site was cleaned, the soldiers debriefed, and the bodies and the craft removed, silence fell. It would not be broken for almost forty-five years.

APPENDIX IV

"PENTAGON ADMITS IT HAS UFO DEBRIS," BY ANTHONY BRAGALIA

A stunning admission by the US government that it possesses UFO debris was recently made in response to a Freedom of Information Act (FOIA) request filed over three years ago by this author. In a reply letter, the US Defense Intelligence Agency (DIA) has ended decades of speculation by verifying that UFO material has indeed been recovered. Now officially referred to as UAP (Unidentified Aerial Phenomena) rather than UFOs, some this material was placed with a defense contractor for analysis and storage in "specialized facilities." Incredibly, part of the information released discusses material with shape recovery properties, much like the "memory metal" debris found fallen at the Roswell UFO crash in 1947.

And based on the documentation received, it appears that the retrieved debris exhibits other extraordinary capabilities. In addition to "remembering" their original form when bent or crushed, some of these futuristic materials have the potential to make things invisible, "compress" electromagnetic energy, and even slow down the speed of light. Although much of the reports' details are redacted, what can be gleaned is that these technologies represent a literal quantum leap beyond the properties of all existing material known to man.

Finally, the whereabouts of the UFO debris held by the contractor is unknown. Some months ago they laid off their employees and ceased operations. Former company officials, when reached, refuse comment.

THE FOIA REQUEST

The original 2017 FOIA request made to the DIA asks for the physical descriptions, properties, and composition of UFO/UAP material held by the government and its contractor. It is unambiguous in its meaning. It refers to UFO/UAP material and "physical debris recovered by personnel of the Department of Defense as residue, flotsam, shot-off material or crashed UAPs or unidentified flying objects."

In its reply, the DIA, amazingly, agrees that it has documents responsive to my request on recovered UFO debris and its analysis, the program under which it was administered, **AATIP** (the **Advanced Aerospace Threat Identification Program**), and that their defense contractor (**Bigelow Aerospace** in Las Vegas, Nevada) has stored the material. They also provide some reports related to the possible applications of the studied material.

The delays and excuses offered up during the years awaiting the FOIA response were many. They include that the records search involved another agency that needed to be consulted; that I had been emailing my former FOIA Officer though she'd left months ago and never received my follow-up communication; that my case had been reassigned to the DIA's FOIA Chief, Steve Tumiski. After a lawsuit was threatened with the retention of counsel, the DIA somehow, after so very long, acted to respond.

THE RELATION TO ROSWELL

The information received includes several pages of what are called "advanced technical reports" on **nitinol**—a shape recovery alloy. Nitinol has similar properties to the "memory metal" found fallen as debris at the Roswell, New Mexico, UFO crash of 1947. The pages found within the FOIA reply on **nitinol** explore its potential for integrating it into the human body for the improvement of health. Well over forty witnesses to Roswell have mentioned a metal-like material that could "remember itself" when deformed or folded, returning to its original state seamlessly and instantly. In a series of investigative articles appearing on the UFO Explorations website article archive section, the case is credibly made that nitinol represents our first attempts made to replicate a material of the craft's construction. The Roswell debris was flown to Wright Field in Ohio after the crash. Some months later Wright granted a contract to Battelle Memorial Institute to begin phase diagrams (recipes) for mixing nickel and ultra-high-purity titanium, required to make memory metal. General Arthur Exon, Base Commander at Wright Patterson in the 1960s, who flew over the crash site in 1947, spoke to researcher Kevin Randle on tape. He stated that it was his understanding that some of the wreckage being tested was comprised in part of "specially processed" titanium.

And in September of 1947, two months after the Roswell crash, General George Shulgen of Air Intelligence described "the materials of construction" of the flying saucers as potentially being made of *"composite or sandwich construction utilizing various combinations of metal and plastics."* Some of the Roswell crash debris was said to display metal-plastic like appearance and characteristics. The DIA sponsored reports received mention of a highly engineered material called "metamaterial" (see below) as comprised of *"composite media."* Metamaterial can be *layered with metal and plastics.*

Throughout the received FOIA documents, mention is made of potential use of some of the materials in "advanced aerospace platforms." Reference is made to desired material characteristics such as being extremely lightweight and tough, like the characteristics of the debris found at the Roswell crash.

OTHER MATERIAL

Extensive reference is also made to the study and application of what the tests called **metamaterial*** including that some of this material:

- **Can be used to "slow down light" and even "bring light to a complete standstill"** (thus implying the ability to manipulate the speed of light (electromagnetic energy traveling at 186,000 miles per second)

- **May induce invisibility by manipulating refraction and reflectivity and increasing light absorption. The reports use terms like "optical isolation" and "transparency" and make reference to a "MetaMirror" technology** (thus implying the ability to make something unable to be seen and/or picked up by radar, scanning, or imaging)

- **Has the interesting ability to "compress electromagnetic energy"** (Such condensing can make information and energy storage smaller and its transfer faster by reducing volume.)

- **Exhibits a particular "tunable resonance" which was likely "determined during the fabrication."** (The phrase "tunable resonance" refers to a vibration of large amplitude from a small stimulus. Very recent research shows potential in the field of "energy harvesting" technologies, or pulling energy from the environment for low-power electronic devices.)

There seem to be many exciting, but far in the future, technical applications of these materials that will change the way our lives are lived.

Portions of the reports center on next-generation **amorphous metals** (also called "liquid metal" or "metallic glass"), which are novel engineering alloys with disordered atomic-scale structure. Metal is crystalline in its solid state, which means it has a highly ordered arrangement of atoms. Amorphous metals though are crazily disordered—a metal-like material so unique that it is believed it may one day replace plastic and metal in many applications. The material is stronger and lighter than any existing metal, can be injection-molded like plastic (no rivets, seams, or joining; smooth like many reported UFOs), and will never corrode or rust. Imagine using a single razor blade for the rest of your life because it stays super sharp forever. A golf club so springy it can drive a ball farther than any pro ever has. An artificial hip implant that performs better than a real hip. A cell phone case that is completely indestructible.

The contract that Bigelow was granted was under the auspices of an official government UFO study (whose existence was revealed by the *New York Times* in 2017). The contract was for Bigelow to construct "specialized modified facilities" to hold the material for testing. The DoD/Pentagon UFO study effort was called the Advanced Aerospace Threat Identification Program (AATIP). It ran for some years (and under more than one name) and, according to informed sources, remains in operation to this day. What is inside these "specially modified facilities" that requires such contracted technical services? What is the nature of this anomalous debris and what are its properties that it requires shielding?

*Metamaterial, a coined and relatively recent word, is believed to be any material engineered to have a property that is not found in naturally occurring materials. Some of these materials appear to be made from assemblies of multiple elements fashioned from composite materials such as metals and plastics. These composite media can be engineered to exhibit unique electromagnetic properties. Made up from subwavelength building blocks (most often based on metals), these metamaterials allow for extreme control over light energy and optical fields, enabling such effects such as negative refraction to be realized.

THE DIA RESPONSE

A letter was sent by the DIA to this author acknowledging the possession of recovered UFO material and the involvement of a defense contractor, granting access to some technical information (while denying most). Scrolling past the letter begins some of the technical reports.

The FOIA official responding, Steve Tumiski, explains that the DIA has identified five documents totaling 154 pages that are responsive to my request. Well over half of the documents have not been released. And such extraterrestrial material study should generate thousands of pages relevant to the request, not just 154. The DIA cites Exemptions 3 and 6 of federal FOIA regulations as reasons that they are compelled to suppress certain of the documents. Exemption 3 allows the withholding of information prohibited from disclosure by federal statute and leaves no discretion on the issue due to potential harm to national security. Exemption 6 generally applies to records that involve personal privacy. It incorporates a balancing test that weighs any privacy interest against the public interest in disclosure.

In the second PDF of documents, provided are several more pages of the 154 pages said to exist on the matter. It is in these documents that we gain the insights on the material as outlined earlier in this article. It is here that we learn of meta-materials, invisibility, slowing the speed of light, compressing electromagnetic energy, implanting memory metal in people, and other remarkable things:

Sample key phrases alluding to technologies these materials could one day give us can be found within the named pages of two PDF FOIA replies: "*slow down light,*" page 18, page 27, page 6; "*bring light to a complete standstill,*" page 17; "*compress energy,*" page 6; "*Nitinol as a Biomaterial,*" page 19; "*new materials-based perfect absorbers*" of light, page 24; "*novel optical isolators,*" page 6.

WHAT DOCUMENTS WERE WITHHELD?

The information provided in the FOIA response seems to represent reports that are directly relevant to what was learned from the study of the UFO debris, and how insight gained from those studies might be applied in the future, *but does not include a detailing of the found debris itself*.

Disappointingly, the reports do not include much of what was requested such as a physical description and the composition of the material, the origin of the material, and the names of the involved scientists. That remains classified. But *technical pursuit areas derived from the study of those materials* (i.e., invisibility, energy concentration, light speed control, intelligent metal) were, in part, released. The released documents help to inform us of the *potential applications* of the materials, but do not offer deep insight into precisely *what the debris is made of*. They speak of "recent experiments" that "provide new concepts" and of "theoretical developments that might result in new materials." The DIA believes it is being responsive to the FOIA request by acknowledging UFO debris and its storage by Bigelow, and by identifying areas of future applications of these materials without having to actually name responsible parties, of what elements the material is comprised, how it is processed, etc.

BIGELOW AEROSPACE: WHERE IS THE DAMN DEBRIS?

Bigelow Aerospace LLC was the brainchild of Robert Bigelow, a hospitality industry billionaire with a strong interest in things cosmic, the other-dimensional, and the extraterrestrial. Bigelow associated with prominent people such as Senator Harry Reid and former rock star Tom DeLonge. Bigelow has had previous "mystery ventures" such as the nonprofit, now-defunct National Institute for Discovery Science (NIDS.) He was also at one time associated with the famous "Skinwalker Ranch" in Utah where purported paranormal activity has been occurring for years. Bigelow brought a unique blend of money and resources, entrepreneurship, and an interest in paranormal investigation to the fore.

In 1999, Bigelow founded Bigelow Aerospace, developing crewed space complexes and inflatable habitats. Bigelow's facilities operated under great secrecy. According to their still-up website, a warning to all is posted on their home page: "If you have been invited to visit Bigelow Aerospace's facilities, please be aware that you will need a valid driver's license or another form of government-issued ID. Unless authorized by management, please note that cell phones, cameras, flash drives and laptops are not allowed on property." In an aerial overview of the Bigelow facilities near Las Vegas, one can see that all of the windows are blocked, and everything is sealed in the extreme. Are these structures what the Pentagon refers to in its contract with Bigelow as "specially modified facilities" to house physical pieces of UFOs?

The contract that Bigelow was granted under the auspices of an official government UFO study (whose existence was revealed by the *New York Times* in 2017 and ran for several years under different names) was to construct "specialized modified facilities" to hold the material for testing. The DoD/Pentagon UFO study effort was called the Advanced Aerospace Threat Identification Program (AATIP). It ran for some years (and under more than one name including AAWSA) and according to informed sources, remains in operation to this day under yet another name. What is inside these "specially modified facilities" that requires such contracted technical services? What is the nature of this anomalous debris and what are its properties that it requires shielding?

Disturbingly, Bigelow Aerospace suddenly—and with little fanfare—quietly laid off nearly every single one of its eighty-five employees in March of last year. No real reason was given other than it was caused by, according to a company spokesperson, "a perfect storm of problems" including the toll of the 2020/21 pandemic.

The concern, of course, is that with mass layoffs and the ceasing of active operation, **where is the UFO debris now? The current whereabouts of these extraordinarily valuable UFO materials are unknown. Is it still held by Bigelow? If yes, why so, when the company is essentially defunct? Did the UFO material transfer to another custodial company? Was it returned to the Pentagon?** In an effort to find out where the alleged extraterrestrial metal is, former security and legal personnel of Bigelow Aerospace were recently contacted.

ROBERT BIGELOW, HIS SECURITY DIRECTOR, AND HIS LAWYER REFUSE COMMENT

Robert Bigelow himself does not answer my emails to him. Bigelow, now in his late seventies as this is written, seems to limit public discussion of his research. So I tried to contact Mr. Ryan Aslesen, the former Director of Security for Bigelow Aerospace. Before employment at Bigelow, Aslesen held security roles in the US Marine Corps. He also has experience in warehousing and logistics management. Today he is a private security consultant in the Las Vegas area. He also self-publishes science fiction, with at least one book involving a hostile alien race. An Inmail message inquiring where the debris may be was sent through LinkedIn, as well as an email through his website contact page, and an email to his home email address directly. None were answered.

A reach-out to the former general counsel for Bigelow Aerospace, Jon Field JD of Las Vegas, Nevada, now in private practice, yielded some insight. In this instance, I was actually able to speak to a former Bigelow insider. I left a message for him with his receptionist, eliciting his immediate callback. The message left was that the call was relative to Field's former affiliation with Bigelow Aerospace. I told him that I was contacting him to find out if he could explain—now that the company is dissolving—where the UFO debris may now reside? Does he know if it remains in a secure location and has it been transferred to another organization? Field wanted to know whom I was "affiliated" with and absolutely refused to discuss the matter in any way, repeating, "I can't comment on that."

I explained that the company is in dissolution and that as a citizen and taxpayer I have the right to know if the UFO debris given by the government to a private concern for storage is still secure, given that company is going defunct. I said to Field that if Bigelow is no longer the custodian contractor for the debris, we have a right to know who is the current custodian. Field did not disagree, only repeating in lawyerly fashion that he could not say anything.

In exasperation, I simply appealed to him: where can I go from here, can he help me at all? I told him I don't want to bother people about this that shouldn't be. He suggested contacting Robert Bigelow directly. I did not tell him that Bigelow has refused me any communication. When I asked Field who represents the company as acting general counsel or as retained outside counsel as the company dissolves (that is, who is the successor taking care of last remaining business matters) he said that he had no idea.

I told him that I did not believe him. I cannot accept that he just up and left Bigelow Aerospace without any knowledge of how the legal function of the corporation would continue to operate without him. He said that he did not like the direction of the conversation and we parted.

A TIME FOR [MORE] ANSWERS

What has been learned through this FOIA investigation is indeed historic. There are UFO materials that have been recovered. The materials being studied may one day bring us great advances in technology and other aspects of our daily lives. But the documents are incomplete by the DIA's admission, and the reasons that they cite are weak. Why not disclose from where the material was retrieved and by whom? Why did they redact the names of the authors of these technical papers? Because once identified, we can question them. The other claimed reason for exemption of their identities and what they learned relates to national security and is apparently so great that "there can be no discretion on the matter." I am going to challenge these exemptions that prevent release of requested information by filing with the agency's FOIA Appeals Officer. They must be compelled to tell us more about this extraordinary extraterrestrial hardware.

APPENDIX V

SEALED AFFIDAVIT OF WALTER G. HAUT

SEALED AFFIDAVIT OF
WALTER G. HAUT

(1) My name is Walter G. Haut.

(2) I was born:____June 3, 1922_____.

(3) My address is: ___2801 N. Kentucky, #142, Roswell, New Mexico_____.

(4) I am retired.

(5) In July 1947, I was stationed at the Roswell Army Air Base in Roswell, New Mexico, serving as the base Public Information Officer. I had spent the Fourth of July weekend (Saturday, the 5th, and Sunday, the 6th) at my private residence about 10 miles north of the base which was located south of town.

(6) I was aware that someone had reported the remains of a downed vehicle by midmorning after my return to duty at the base on Monday, July 7. I was aware the Major Jesse A. Marcel, head of intelligence was sent by the base commander, Col. William Blanchard to investigate.

(7) By late in the afternoon that same day, I would learn that additional civilian reports came in regarding a second site just north of Roswell. I would spend the better part of the day attending to my regular duties hearing little if anything more.

(8) On Tuesday morning July 8, I would attend the regularly scheduled staff meeting at 7:30 AM. Besides Blanchard, Marcel, CIC Capt. Sheridan Cavitt, Col. James I. Hopkins, the operations officer, Major Patrick Saunders, the base adjutant, Major Isidore Brown, the personnel officer, Lt. Col. Ulysses S. Nero, the supply officer, and from Carlswell AAF in Fort Worth Texas, Blanchard's boss, Brig. Gen. Roger Ramey, and his chief of staff, Col. Thomas J. DuBose were also in attendance. The main topic of discussion was reported by Marcel and Cavitt regarding an extensive debris field in Lincoln County approx. 75 miles NW of Roswell. A preliminary briefing was provided by Blanchard about the second site approx. 40 miles north of town. Samples of wreckage were passed around the table. It was unlike any material I had or have ever seen in my life. Pieces which resembled metal foil, paper thin yet extremely strong, and pieces with unusual markings along their length were handled from man to man each voicing their opinion. No one was able to identify the crash debris.

(9) One of the main concerns discussed at the meeting was whether we should go public or not with the discovery. Gen. Ramey proposed a plan which I believe originated with his bosses at the pentagon. Attention needed to be diverted from the more important site north of town by acknowledging the other location. Too many civilians were already involved and the press already was informed. I was not completely informed how this would be accomplished.

12/20/2002

(10) At approximately 9:30 Col. Blanchard phoned my office and dictated the press release of having in our possession a flying disc, coming from a ranch northwest of Roswell, and Marcel flying the material to higher headquarters. I was to deliver the news release to radio stations KGFL and KSWS, and newspapers the *Daily Record* and the *Morning Dispatch*.

(11) By the time the news had hit the wire services, my office was inundated with phone calls from around the world. Messages stacked up on my desk and rather than deal with the media concern, Col. Blanchard suggested that I go home and "hide out."

(12) Before leaving the base, Col. Blanchard took me personally to Building 84, a B29 hangar located on the East side of the tarmac. Upon first approaching the building, I observed that it was under heavy guard both outside and inside. Once inside, I was permitted from a safe distance to first observe the object just recovered north of town. It was approx. 12-15 feet in length, not quite as wide, about 6 feet high and more of an egg shape. Lighting was poor but its surface did appear metallic. No windows, portholes, wings, tail section, or landing gear was visible.

(13) Also from a distance, I was able to see a couple of bodies under a canvas tarpaulin. Only the heads extended beyond the covering and I was not able to make out any features. The heads did appear larger than normal and the contour of the canvas over the bodies suggested the size of a 10 year old child. At a later date in Blanchard's office, he would extend his arm about 4 feet above the floor to indicate the height.

(14) I was informed of a temporary morgue set up to accommodate the recovered bodies.

(15) I was informed that the wreckage was not "hot" [radioactive].

(16) Upon his return from Fort Worth, Major Marcel described to me taking pieces of the wreckage to Gen. Ramey's office and after returning from a map room, finding the remains of a weather balloon and radar kite substituted while he was out of the room. Marcel was very upset over this situation. We would not discuss it again.

(17) I would be allowed to make at least one visit to one of the recovery sites during the military cleanup. I would return to the base with some of the wreckage which I would display in my office.

(18) I was aware of two separate teams which would return to each site months later for periodic searches for any remaining evidence.

(19) I am convinced that what I personally observed was some type of craft and its crew from outer space.

(20) I have not been paid nor given anything of value to make this statement, and it is the truth to the best of my recollection. THIS STATEMENT IS TO REMAIN SEALED AND SECURED UNTIL THE TIME OF MY DEATH AND AT WHICH TIME MY SURVIVING FAMILY WILL DETERMINE ITS DISPOSITION.

Signature, Walter G. Haut

Signature Witnessed by:

Name

Date 12/26/02

OFFICIAL SEAL
BEVERLEE J. MORGAN
NOTARY PUBLIC
STATE OF NEW MEXICO
My Commission Expires

12/20/2002

(Photo courtesy of Don Schmitt/IUFOM&RC.)

NOTES

PREFACE

1. *www.billboard.com.*
2. "Movies of the Year: The Big List, 1915-1949." *www.theendofcinema .blogspot.com.*

CHAPTER I

1. The term *Roswell Incident,* used most often to describe the alleged crash, recovery, and cover-up of a UFO and its crew near the town of Roswell, New Mexico, in 1947, is derived from the title of the first book ever published on the subject: *The Roswell Incident* by Charles Berlitz and William L. Moore, Grosset & Dunlap: New York (1980).
2. United States General Accounting Office, *Results of a Search for Records Concerning the 1947 Crash near Roswell, New Mexico,* Pub. No. GAO /NSIAD-95-187, Washington, DC: GAO, July 1995.
3. The Air Force's Project Mogul balloon hypothesis is articulated in detail in its massive publication, *The Roswell Report: Fact versus Fiction in the New Mexico Desert,* by Richard Weaver, Washington, DC: U.S. Government Printing Office, 1995 (a 23-page Executive Summary was published in 1994). Its time compression/dummies-from-the-sky hypothesis can be found in *The Roswell Report: Case Closed* by James McAndrew, Washington, DC: U.S. Government Printing Office/Barnes & Noble, Inc., 1997.
4. Broad, "Wreckage."

CHAPTER 2

1. Sheridan Cavitt, Lt. Col. USAFR statements to Col. Richard Weaver. In our initial meetings with Cavitt, he denied repeatedly that he was at Roswell at the time of the incident. May 24, 1994.

2. Jesse A. Marcel, Lt. Col. USAFR. This and similar statements were made until his death in 1986. Marcel would also emphasize that there was much more to the story and that he was not fully speaking to "protect the country."

3. Frank Joyce, personal interviews 1998-2002.

4. Ibid.

5. Cavitt interview with Weaver. On different meetings with Cavitt, we secretly recorded our conversations. Every detail Cavitt confided to Weaver was never mentioned to us. In stark contrast, Cavitt denied everything he acknowledged to Weaver during their interview on May 24, 1994.

6. Cavitt to Weaver in total contradiction of his testimony to us. As we were departing for the last time from the Cavitt residence in Sequim, Washington, June 1994, Cavitt's wife Mary remarked, "Tell the boys about the colonel from the Pentagon who came to see you last month." To which her husband snapped, "Yes, and I told him just as I have been telling you, I wasn't there!" As his sworn statement to Weaver clearly demonstrates, this was not even close to being true.

7. Weaver, *The Roswell Report*, 1994.

8. Ibid.

9. Ibid.

10. Ibid.

11. Lewis Rickett, personal interviews, 1989-1992; videotaped interview by coauthor Schmitt, 1990.

12. Ibid.

13. Ibid.

14. Ibid.

15. Ibid.

16. Ibid.

17. Transcript of interview of Sheridan Cavitt by Richard Weaver, May 24, 1994.

18. Rickett personal interviews, 1989-1992.

19. Interview of Cavitt by Weaver, 1994.

20. Ibid.

21. Rickett, personal interviews, 1989-1992.

22. Ibid.

23. Ibid.

24. Ibid.

25. Ibid.

26. Karl T. Pflock, *Roswell in Perspective*, 1994. During the interview between Cavitt and Weaver, Cavitt asks, "You've heard of Pflock?" Weaver responds, "I know who he is, yeh." Cavitt adds, "He's our chief debunker. I lean toward him." At the time of the Cavitt/Weaver interview, Pflock had yet to publicly endorse the Mogul balloon theory.

27. Cavitt, personal interview, William L. Moore, 1985.

28. Mary Cavitt, personal interview, Moore, 1985.

29. Rickett, telephone interview, 1992.

30. Julie Rees Rudzinski, personal and telephone interviews, 2013.

31. Doyle Rees, Col. USAFR, letter to Cavitt, 1992.

CHAPTER 3

1   Bill Brazel, personal interviews, 1989-2003.

2. Loretta Proctor, personal interviews, 1989-1990, 1994-1995.

3. Sydney Wright, personal interview, 1998.

4. Hope Baldra, personal interview, 1998.

5. Budd Eppers, personal interview, 1999.

6. Truman Pierce, personal interview, 1999.

7. Glaze Sacra, personal interview, 1999.

8. Danny Boswell, personal interviews, 1999-2000.

9. Shirley Brazel, personal interview, 1990.

10. Bill Brazel, personal interviews, 1989-2003.

11. Ibid.

12. Charlie Schmid, personal interviews, 1991, 1993.

13. Paul Price, personal interview, 2006.

14. Robert Smith, personal interview, 1992.

15. Trini Chavez, personal interview, 2005.

16. Loretta Proctor, personal interview, 1994.

17. Budd Payne, personal interviews, 1990-1991.

18. Fawn Fritz, personal interviews, 2005-2006.

19. Jeff Wells, personal interviews, 1989-1990.

20. Robert Smith, personal interview, 1992.

21. L. D. Sparks, personal interview, 2000.

22. Jack Rodden, personal interview, 1990.

23. Leroy Lang, personal interview, 1998.

24. Hope Baldra, personal interview, 1998.

CHAPTER 4

1. Berlitz and Moore, *The Roswell Incident.*

2. In-person interviews conducted over the years with Mack Brazel's son, Bill Brazel, his daughter Bessie Brazel Schreiber, and his granddaughter Fawn Fritz.

3. Frank Joyce, personal interview, May 1998.

4. Jud Roberts, personal interviews, 1991, 1993.

5. Frank Joyce, personal interviews, 1989-2006. According to Joyce, the very next morning after the weather balloon press conference in Fort Worth, MPs arrived at KGFL and went through every room and every desk drawer, searching for any reference contrary to the balloon explanation.

6. General Michael Rexrold, personal interviews, 1990-1991. In 1947, Rexrold was the personal assistant to Senator Dennis Chavez. Rexrold told us that the senator never mentioned the Roswell Incident to him, but did make an unscheduled trip back to New Mexico after the holiday weekend.

7. Lydia Sleppy, personal interview, 1991.

8. Merle Tucker, personal interview, 1989. According to Tucker, he knew that Johnny McBoyle had tried to get out to the site just north of town and that he had been intercepted by the military. Once McBoyle returned to Roswell, the sheriff went to see him and warned him not to ever talk about it again. Tucker had no contact with either of the New Mexican senators or anyone else. He did hear that someone "official" had gone into his Roswell station and cleaned out all the paper about the event. Merle Tucker was the owner of KSWS and KOAT.

9. Lydia Sleppy, personal interview, 1991.

10. John McBoyle, telephone interview, 1990.

11. Frank Joyce, personal interviews, 1989-2006.

CHAPTER 5

1. Robin Adair, personal interview, 1993.

2. Ibid.

3. Ibid.

4. "Harassed Rancher," *Roswell Daily Record.*

5. Ibid.

6. Ibid.

7. Randle and Schmitt, *UFO Crash at Roswell.*

8. Thurber, "Allan Grant."

9. See the Allan Grant website that features the pictures he took for *Life* magazine, *www.allangrant.com.*

10. Mrs. Allan Grant, interviews with Anthony Bragalia via email, 2007-2008; interviews with Tom Carey via email, 2008.

11. Ibid.

12. Ibid.

CHAPTER 6

1. *Roswell,* Showtime/Viacom, 1994.

2. Jud Roberts, personal interview, 1991.

3. "Harassed Rancher," *Roswell Daily Record.*

4. Bill Brazel, personal interviews, 1989-2003.

5. Joe Brazel, personal interview, 2004.

6. Ibid.

7. Bill Brazel, personal interviews, 1989-2003.

8. Shirley Brazel, personal interviews, 1989-1991.

9. Bessie Brazel Schreiber, personal interviews, 1990-1991, 2000-2001.

10. Bill Brazel, personal interviews, 1989-2003.

11. Frank Joyce, personal interviews, 1989-2006.

12. Ernest Lueras, personal interview, 2000.

13. Bill Brazel, personal interview, 1995. It was later discovered that Emerson Armstrong was in charge of UFO investigations for the southwestern part of the United States out of Fort Bliss in El Paso, Texas, in 1949. Even after providing him with documentation to that effect, he denied it was him and claimed he never heard of the Roswell Incident.

14. Bobby Wade, personal interviews, 1998, 2000.

15. Fawn Fritz, personal interviews, 2005-2006.

16. Bert Schulz, personal interview, 2008.

17. Michelle Penn, telephone interview with Anthony Bragalia, 2008.

18. Howard Scoggin, telephone interview, 1999.

19. Bob Wolf, personal interviews, 1992, 1999.

20. Fawn Fritz, personal interviews, 2005-2006.

CHAPTER 7

1. Cory Derek, telephone interview, 2008.

2. Geraldine Perkins, in-person interview, 2000. Also in-person interviews (2000–2008) with Mack Brazel's grandson and granddaughter-in-law, Joe and Beverly Brazel, who have confirmed that their grandfather would have indeed contacted his boss, J. B. Foster, within short order about the crash on the latter's property; Brazel's neighbor, Juanita Sultemeier, concurred (in-person interviews, 2006, 2008): "Mack's job was at stake if he didn't report the situation to his boss."

3. JoAnn Purdie, telephone interview, 2008.

4. Cory Derek, telephone interview, 2008.

5. JoAnn Purdie, telephone interview, 2008.

6. Ibid.

7. Headquarters United States Air Force. Colonel Richard L. Weaver. *Report of Air Force Research regarding the "Roswell Incident."* Also, telephone interview with Bessie Brazel Schreiber, 1989, and in-person interview with Bessie Brazel Schreiber, 1999.

8. Bessie Brazel Schreiber, telephone interviews, 2001-2008.

9. Bill Brazel, telephone interview, 1989.

10. Bessie Brazel Schreiber, personal interview, 1999.

11. JoAnn Purdie, telephone interview, 2008.

CHAPTER 8

1. Jesse Marcel Sr., transcript of personal interview conducted by Bob Pratt, December 1979.

2. Robert Slusher, personal interviews, 1991, 2002.

3. Jesse Marcel Jr., personal interview, 1991.

4. Jesse Marcel Sr., transcript of personal interview conducted by Bob Pratt, December 1979.

5. Lewis Rickett, personal interviews, 1991-1992.

6. Mary Cavitt, personal interview, 1991.

7. Ibid.

8. Herschel Grice, telephone interview, 2002.

9. Elaine Vegh, personal interview, 1990.

10. Elaine Vegh, telephone interview with Anthony Bragalia, 2007.

11. Morris, "UFO whistleblower."

12. Jesse Marcel Sr., transcript of personal interview conducted by Bob Pratt, December 1979.

CHAPTER 9

1. We believe that the lack of a paper trail in the public record for Montoya's whereabouts on that day was done deliberately for political reasons to keep his name from being associated with the Roswell Incident, as well as for convenience in order to prevent the inundation of phone calls to his office that would surely follow.

2. David Rudiak, *roswellproof.com.*

3. A number of witnesses have testified that "one was alive." Alpha Edwards told our investigation in 1999 that her father had been a civilian working in the hangar area of the base when the "little men" arrived. He told her, "Everybody was running around, and a group of men carried one of them

right past me and into the hangar. At first, I thought it was a small child because of its size. Then, after getting a quick look at it, I realized from the size and shape of its head and eyes that it wasn't from 'around here,' that it wasn't human. It was alive but appeared to be injured. I kept wondering why, if it was injured, they weren't taking it over to the hospital." This anecdotal account tracks with Joseph Montoya's claim of seeing one that was alive but injured inside the hangar lying on a table brought over from the mess hall.

4. Pete and Mary Anaya, personal interview, September 2002.

5. Headquartered at Wright-Patterson Air Force Base in Dayton, Ohio, Project Blue Book was the Air Force's official investigative unit for reports of UFOs. It was in existence from 1951 to 1969, during which it investigated in excess of 12,000 reported UFO cases, of which slightly more than 700 were designated as "unknown." Project Blue Book was shut down in 1969 at the conclusion of a three-year study of the UFO phenomenon by the University of Colorado. The Air Force no longer "officially" investigates UFOs.

CHAPTER 10

1. Thomas J. DuBose, personal interview, 1991.

2. Ibid.

3. Weisgall, *Operation Crossroads*. Ramey retired from the Air Force as director of the Air Defense Command in 1957 as a three-star general. It remains curious that as a lieutenant general with a most impressive military history, including having an Air Force base named after him, Ramey's biography is not included on the USAF website, which does provide backgrounds on more than 2,000 other high-ranking officers.

4. As reported in the evening edition of the Los Angeles *Herald Express* on July 8, 1947, from an International News Service news dispatch; information provided by David Rudiak via personal correspondence; see also his website, *roswellproofcom.*

5. "General Ramey," *Roswell Daily Record.*

6. FBI telex of July 8, 1947, from FBI Agent Percy Wyly of the FBI's Dallas office to the FBI office in Cincinnati, and to FBI Director J. Edgar Hoover; the telex also suggests that the wreckage received at Wright Field was something more than General Ramey's claimed weather balloon.

7. Joe Briley, telephone interviews, 1989, 1990, and 2001; Lieutenant Colonel Briley was on Colonel William Blanchard's immediate staff, as the chief operations officer at the RAAF in 1947; in 2005, we contacted Briley again, this time for an in-person interview. He turned down our request. In 2007, as a last resort, we sent him a copy of the first edition of this book with a request to call us with comments after reading it. He never did. The information about Colonel Blanchard's temporarily moving his quarters to one of the EM barracks comes from two enlistees whose names are on file. There is also a story that, when word got around, Blanchard moved his quarters again—to the base jail!

8. See Walter Haut affidavit in this book.

9. From a July 9, 1947, Associated Press wire story that appeared in the *Albuquerque Journal* and the *San Antonio Express;* source information provided by David Rudiak (*roswellproof com*).

10. Jason Kellahin. Affidavit, signed September 20, 1993.

11. Newspaper story from the *Santa Fe New Mexican,* July 9, 1947; source information from David Rudiak (*roswellproof com*).

12. "Balloon Demonstrations." See David Rudiak's website devoted to the Roswell Incident, *roswellproofcom.*

CHAPTER II

1. Patricia Rice, telephone interview, June 3, 2000 (and quoted in the *Dallas Morning News,* July 6, 1997).

2. Earl V. Fulford, telephone interviews, 2005–2006.

3. The wording of the press release strongly suggested that the 509th Bomb Group had in its possession the previous day, July 7, 1947 (even before Major Marcel had returned from the debris field on the Foster ranch in Lincoln County), enough evidence to conclude that the recovery then in progress involved an extraterrestrial event. It also suggests that the military was trying to divert the attention of the Roswell citizenry away from the crash site just north of town to a site farther away, nearer to Corona.

4. Richard Talbert, telephone interviews, 2003–2004.

5. May Rich (Rich's widow), telephone interview, 2004.

6. Paul McFerrin, telephone interview, 2003.

7. Jobie MacPherson, telephone interviews, 2001–2002.

8. Michael Menagh and Rolland Menagh Jr., telephone interviews, 2005.

9. Tilley and Tilley, *Exposé*.

10. Ibid.

11. Chester P. Barton, telephone interview with Joseph Stefula, 1995.

12. Earl Fulford, telephone interviews, 2005–2006.

13. Harvie L. Davis, telephone interview, 2005.

14. John Bunch, telephone interview, 2006.

15. Eugene C. Heines, telephone interviews, 2005.

16. Earl Fulford, telephone interviews, 2005–2006.

17. George Houck, telephone interview, 2005.

18. Earl Fulford, telephone interviews, 2005–2006.

19. Frank and Mary Martinez, personal interview, 2005.

20. There were in fact several courts-martial handed down at the time: Earl D. Downs was a PFC with the 390th ASS in charge of a detachment of airmen guarding the hangar containing the wreckage—he was court-martialed and thrown in the brig for several months, along with the men of his detachment, because one of them had, unknown to him, pocketed a small piece of wreckage from the hangar they were guarding.

21. Harry Girard, telephone interview, 2005.

22. Steve Whalen Jr., telephone interview, 2005.

23. Earl Fulford, telephone interviews, 2005–2006. Fulford's account receives corroboration from the former Staff Sergeant Milton C. Sprouse, who was a B-29 crew chief in the 830th Bomb Squadron in 1947. In a telephone interview in 2001, the seventy-four-year-old Sprouse told our investigation that he and his crew had just returned to Roswell from Florida on *Daves, Dream* the evening of July 7. Sprouse said that the base was buzzing with talk about the crash. A week or so later, a few of Sprouse's buddies told him that they had been working late on one of the Silverplates the previous evening. They told Sprouse that they witnessed a C-54 transport pull up in front of the hangar (Hangar P-3), whereupon it was loaded with pieces of wreckage. They especially remembered one large piece that was loaded, but they

couldn't get a closer look because of the guards posted around the hangar. "The rumor was that General Ramey was on the plane." In the morning, the plane was gone, and the hangar was empty.

24. William C. Ennis, personal interviews, 1992, 2008.

25. George Newling, telephone interview, 1999; in-person interviews, 2000, 2002; see also Newling's account in Tilley and Tilley, *Exposé*.

CHAPTER 12

1. Thomas J. DuBose, personal interview, 1991.

2. Julian P. Anderson, telephone interview, 2008.

3. Robert J. Barrowclough, personal interview, 2000.

4. Robert R. Porter, personal interview, 1991.

5. Walter G. Haut, personal interviews, 1989–2005.

6. James B. Johnson, personal interview, 1990.

7. Robert Porter, personal interview, 1991.

8. Robert B. Shirkey, personal interviews, 1990–2002.

9. Ibid.

10. Robert Porter, personal interview, 1991.

11. Ibid.

12. The Irving Newton photo is not shown.

CHAPTER 13

1. Rosemary McManus, personal interview, 1994.

2. *Atomic Blast*, February 21, 1947.

3. Patricia Bush, personal interviews, 2004, 2006.

4. Glenn Dennis, personal interview, 1990.

5. E. M. Hall, personal interview, 1993.

6. George Bush, personal interview, 1990.

7. Ibid.

8. Ibid.

9. Ibid.

10. Jean Bush Overton, telephone interview, 1992.

11. Patricia Bush, personal interviews, 2004–2006.

12. Ibid.

13. Victor Golubic, quoted in Pflock, *Roswell: Inconvenient Facts.*

CHAPTER 14

1. Berlitz and Moore, *The Roswell Incident.*

2. Randle and Schmitt, *UFO Crash.*

3. Interview of the late Melvin Brown's daughter, Beverly Bean, in *Recollections of Roswell, Part II.*

4. Interview of the late Pappy Henderson's wife, Sappho Henderson, in *Recollections of Roswell, Part II.*

5. Interview of Glenn Dennis, in *Recollections of Roswell, Part II.*

6. Randle and Schmitt, *The Truth.*

7. The late Frank Kaufmann, a former personnel clerk at the Roswell base and later a thirty-year member of the Roswell Chamber of Commerce, claimed to have been a member of a super-secret ensemble, "The Nine." According to Kaufmann, the group was responsible for containing highly classified accidents and operations, such as that presented by the Roswell UFO recovery. Before his death, Kaufmann admitted to us that he made up the Roswell crash site location. Asked why he didn't say anything previously when so much attention was being paid to it by the media, Kaufmann simply said, "It got out of control."

8. Loretta Proctor, personal interview, 1996. This site, known as the "Dee Proctor Body Site" is the *someplace else* where Mack Brazel told Frank Joyce that he had found bodies of "little people" during his first conversation with Joyce on July 6, 1947.

9. Edward Sain, telephone interview, 2005.

10. Steven Sain, telephone interview, 2005.

11. Mrs. Raymond Van Why, telephone interview, 2005. Mrs. Van Why's account of her husband's observation of a "disc," and Edward Sain's identification of Van Why as the one who guarded the bodies with him, suggest that Van Why had been a guard at two sites: the impact site and the Dee Proctor body site.

12. Mrs. Wallace, in-person interview for our investigation, conducted by Dennis Balthaser, 1999.

13. From Frederick Benthal's notarized statement, signed May 5, 1993; see also Benthal's videotaped interview in *UFO Secret: The Roswell Crash Update 2000.*

14. William J. Warnke, telephone interview, 2006.

15. Monte Dalton, telephone interview, 2005.

16. Pflock, *Roswell: Inconvenient Facts.* Dennis also confirmed this account to us several times since 1998.

17. Arthur Fluery's widow, telephone interview, 2005. Paul Camerato's widow confirmed her late husband's presence as an ambulance driver in Roswell at the time of the incident, but could provide no additional information.

18. Adam Dutchover, telephone interview, 2008.

19. Garner Mason, telephone interview, 2008.

20. Rex Alcorn, telephone interview, 2008.

21. Clifford Butts and William Burkstaller, telephone interviews, 2008.

22. Richard L. Bean, Esq., personal interview, 2004.

23. L. M. Hall, signed affidavit, September 15, 1993.

24. Beverly Otto, telephone interview, 2008.

25. Headquarters United States Air Force. Captain James McAndrew. *The Roswell Report: Case Closed.*

26. Bob Wolf, personal interview, 1999.

27. LeRoy Lang, personal interview, 1999.

28. Mr. and Mrs. Eleazar Benavidez, personal interviews, 2003–2006.

29. Mrs. David Ackroyd, telephone interview, 2004.

30. The Army Air Corps became a subordinate element of the Army Air Forces on June 20, 1941, and continued to exist as a combatant arm of the Army Air Forces until both entities unified as a separate branch of the military, the U.S. Air Force, on September 18, 1947.

31. Joseph Toth, telephone interview, 2008.

CHAPTER 15

1. Stringfield, *The UFO Crash/Retrieval Syndrome.* Also, personal interviews, 1990–1992.

2. Ibid.

3. Ibid.

4. Ibid.

5. Beverly Bean, personal interview, 1991.

6. Leonard Stringfield, personal interviews, 1990–1992.

7. Son (anonymity requested) of Dr. Foster's housekeeper (anonymity requested), personal interview, 1992.

8. Frankie Dwyer Rowe, personal interviews, 2005–2006.

9. The phrase *victims of the wreck* comes from the famed "Ramey memo," a telex that General Roger M. Ramey was photographed holding in his hand by a photographer from the *Fort Worth Star Telegram* newspaper during Ramey's "It was a weather balloon" press conference on July 8, 1947. Most Roswell researchers believe this phrase refers to the unfortunate crew from the downed ("wrecked") UFO.

10. Daughter (anonymity requested), personal interview, 1992.

CHAPTER 16

1. The witness list includes first-, second-, and thirdhand witnesses. Legal proceedings consider anything other than firsthand, eyewitness testimony as "hearsay." However, if enough "hearsay" testimony is gathered that supports the case being presented, it can form the basis for a powerful corroboration of the case.

2. During WWII, Native Americans were often utilized in situations that required the implementation of extreme security, the thought being that Native Americans could be trusted to keep their mouths shut. At the RAAF in 1947, there was a small security detachment of Native Americans utilized in such highly sensitive situations.

3. Edward Harrison, telephone interview, 2005.

4. The 393rd Bomb Squadron was the original bomb squadron of the 509th Composite Group when it was formed in 1944. It wasn't until after the war, when the 509th deployed to Roswell, that two additional bomb squadrons, the 715th and the 830th, joined the Bomb Group.

5. During the atomic bombing of Hiroshima, the *Necessary Evil* was responsible for photographing the event for posterity, while the *Straight Flush* was responsible for monitoring the weather over the target.

6. In 1947, a civilian by the name of Nancy Havins became the first female flight mechanic at the RAAF. She passed away in 2002, but our investigation interviewed her three sons, who told us that their mother had been a "fabricator" at the base, whose job was to construct wooden crates to ship engines, parts, and so on from the base to other locations. Her sons stated that she had gone to the crash site in 1947 to see what size and how many boxes would be needed to transport the UFO wreckage and presumably the bodies.

CHAPTER 17

1. Copy of the July 9, 1947, flight log of the *Straight Flush* supplied by Lloyd Thompson.
2. Unless otherwise stated in the text, the details of the July 9, 1947, flight of the *Straight Flush* are based upon telephone conversations between Lloyd Thompson and Len Stringfield; between Lloyd Thompson and Don Schmitt; between Lloyd Thompson's son, Lowell Thompson, and Tom Carey; an audiotaped presentation by Lloyd Thompson to a local Boise, Idaho, UFO group in 1992; in-person and telephone interviews of Robert Slusher with Don Schmitt and Tom Carey; and telephone interviews of Arthur Osepchook by Tom Carey in 2003 and 2006.

CHAPTER 18

1. Fawn Fritz, personal interview, 2006; Ardeth Vandercook, telephone interview, 2000.
2. Frank Joyce, personal interviews, 1998–2002.
3. Scroggins's widows, telephone interviews, 1999–2000.
4. Frankie Dwyer Rowe, personal interviews throughout many years.
5. J. C. Smith, separate telephone interviews with Tom Carey and Anthony Bragalia, 2004 and 2007.
6. Sue Farnsworth Bennett, personal and telephone interviews, 2008.
7. George Cisneros, telephone interviews, 2008.
8. Sue Farnsworth Bennett, personal and telephone interviews, 2008.
9. Douglas Loveridge, telephone interviews, 2007-2008.
10. Ibid.

11. AP wire story that ran in the *Albuquerque Journal* and *San Antonio Express*, unattributed but likely authored by Jason Kellahin. This information was given to us by Roswell researcher David Rudiak (see also *roswellproof.com*).

12. Randle and Schmitt. *The Truth about the UFO Crash at Roswell*.

13. Ibid.

14. Barbara Dugger, granddaughter of George and Inez Wilcox, videotaped interview in *Recollections of Roswell, Part II*.

15. Michelle Penn, interview with Anthony Bragalia, 2008.

CHAPTER 19

1. History of Timken Company, Timken Company website. Founded in 1899, Henry, H. H., & William Timken, Timken Roller Bearing Company.

2. Native of Roswell Thomas J. Brookshier, NFL all-pro defensive cornerback for the Philadelphia Eagles and legendary sportscaster with CBS, told co-author Carey about witnessing a piece of memory metal at Tyner's Welding Shop at the time of the reported crash.

3. Personal interviews conducted by William E. Jones and Irena McCammon, *Ohio UFO Notebook*, 1994.

4. Ed Balint, *The Repository: Canton's Close Encounter*, August 22, 2010.

5. Roswell City Directory, 1947 edition.

6. James Wood, telephone interview, 2012.

7. Frank Vega, personal interviews, 2012, 2013.

8. James Wood, telephone interview, 2012.

9. Frank Vega, personal interviews, 2012, 2013. Numerous other accounts on file about Roswell Army Air Field personnel who were immediately transferred after incident, including outside enlisted men from Fort Worth, White Sands, Wright Field, and Fitzsimmons with special assignments to Roswell at the time of the incident.

10. Ibid.

11. All quotes from this paragraph from James Wood, telephone interview, 2012.

12. Ibid.

13. Ibid.

14. Ibid.

15. Special thank you to Dr. Michael Swords, professor emeritus, natural history department, University of Michigan at Kalamazoo.

CHAPTER 20

1. Moore, "The Roswell Investigation."

2. Ibid.

3. General William H. Blanchard's daughter Dale, quoted in Lawrence, "Did a UFO Really Crash at Roswell?"

4. Art McQuiddy, signed affidavit, October 19, 1993.

5. William Brainerd, telephone interview, 1993.

6. Hastings (under General Ramey), *UFOs and Nukes.*

7. Marion Brimberry, telephone interview, 2000.

8. Elaine Shepard, telephone interviews, 1997–1998; also telephone interviews with UFO documentarian Sam Sherman, who provided the lead, 1997.

9. Arthur Exon, personal interviews, 1990–1991, 1994, 2000.

CHAPTER 21

1. Richelson, *U.S. Intelligence Community.*

2. FBI Dallas Bureau Transmission, July 8, 1947.

3. Arthur Exon, personal interviews, 1990–1991, 1994, 2000.

4. Ibid.

5. Ibid.

6. Ibid.

7. Ibid.

8. Joseph Briley, personal interviews, 1992, 2000, 2006.

9. Arthur Exon, personal interviews, 1990–1991, 1994, 2000.

10. Ibid.

11. El Paso radio interview of General Roger Ramey in which it was stated that "...only people in Kansas didn't report any flying saucers because it was a dry state." Prior to the Roswell Incident, Ramey is quoted as referring to such reports as "Buck Rogers stuff." Ramey said, "People were probably seeing heat waves or misidentifying distant jet planes." Associated Press/United Press International.

12. Richelson, *U.S. Intelligence Community*.

13. Earl L. Zimmerman, affidavit, November 2, 1993; also telephone interview, 1995.

14. Ibid.

15. Benjamin Games's story was first reported by Billy Cox in the *Sarasota Herald Tribune;* Cox subsequently contacted our investigation and Anthony Bragalia, who lives near Games; each then conducted separate interviews of Games in 2008, ours by telephone, and Bragalia's in person.

CHAPTER 22

1. James Parker, telephone interview, 1990.
2. Ibid.
3. Charles McGee, telephone interviews, 1992-1993.
4. Ibid.
5. Rogene Com Cordes, separate telephone interviews with Tom Carey and Anthony Bragalia, 2006–2008.
6. Charles McGee, 1992–1993.
7. Ibid.
8. Walter Haut, personal interview, 2000.
9. Bill Brazel, personal interview, 1989.
10. Don Bogle, telephone interview, 1989.
11. Pat Flanary, telephone interview, 1989.
12. James Parker, telephone interview, 1989.

CHAPTER 23

1. Edwin Easley, telephone interview, 1990.
2. Nancy Strickland, personal interviews, 1990, 1992, 2002.
3. Dr. Harold Granek, personal interviews, 1991, 2002.
4. Family of Patrick Saunders, personal interviews, 1991, 1995.
5. Schiff, "News Release."

6. Sarah Mounce, telephone interview, 2000.

7. Wanda Lida, telephone interview, 2000.

8. Larry Rowlette, personal interview, 2005.

9. Carlene Rowlette Green, personal interview, 2005.

10. Daughter of Harry Telesco (name on file), telephone interview, 2007.

11. Family member of James Saine (name on file), telephone interview, 2008.

12. Beverly Bean, personal interview, 1991.

13. Mary Ann Gardner, personal interview, 1990.

14. Barbara Dugger, personal interview, 1991.

15. Blanche Wahnee, Wanda Wahnee Priddy, and Meyers Wahnee Jr. (Meyers Wahnee's children), telephone interviews, 2005.

16. Mark Magruder, telephone interviews 2002, 2008.

17. John Kromschroeder, videotaped interview in *Recollections of Roswell, Part II.*

18. Herschel Grice, telephone interview, 2003.

19. Birnes, "Squiggly."

20. Jack Rodden Jr., videotaped interview in *Recollections of Roswell, Part II.*

21. Sheridan Cavitt, personal interviews, 1990–1994.

22. Mary Cavitt, personal interviews, 1990–2000.

23. Joseph Cavitt, personal interview, 1998.

24. Cavitt personal physician (anonymity requested), personal interview, 2001.

25. Kenneth Compton, telephone interviews, 2008.

26. Name on file, telephone interview, 2001.

27. "Major General Robert B. Landry," *Air Force Link.* Official website of the United States Air Force, 2008. *af.mil.*

28. "President Harry S. Truman: 33rd President, April 12, 1945–January 20, 1953," by Grant Cameron; from Cameron's website devoted to presidential knowledge of and involvement in the UFO phenomenon, 2008. *www.presidentialufo.com.* (Website no longer available.)

29. Dr. Meg Blackburn Losey, telephone interviews, 2007.

CHAPTER 24

1. The quote is attributed to an unnamed former military officer by the late Karl Pflock in his 2001 book *Roswell: Inconvenient Facts and the Will to Believe.*

2. Lawrence, "Did a UFO Really Crash at Roswell?"

3. Ibid.

4. Proof of this came from an interview that we conducted in 2002 with the seventy-four-year-old Lloyd E. Nelson, a former PFC who clerked for Walter Haut in the RAAF base Public Information Office in 1947. According to Nelson, he remembers Walter Haut coming into the office at the time of the incident with Major Marcel. He had assumed that his boss and Marcel had just come back from the crash site, because each was holding and showing around the office small pieces of wreckage debris. "I was shown an I-beam or two by Marcel and Walter that were very short and had reddish writing on them that I did not recognize. I handled these myself. I also handled a very hard ceramic-type material like flint that appeared to have been broken off something. It was darker on one side and lighter on the other. It really wasn't metal. I also was shown and handled three pieces of very thin, aluminum-like metal, which was very hard. Marcel told me that they had been testing the metal and that all of it was very light and very hard, and that it was unlike anything that anyone had ever seen. Before they left the office, Marcel and Walter told everyone present not to say anything about what they had just seen. I believe that this meeting took place on Friday, July 11, because I remember that I didn't work the next day." In 2000 or 2001, Lloyd Nelson called Walter Haut to try to find out what Walter knew and why he had been so quiet about it over the ensuing years, considering what he knew to be true about Walter. "To my dismay," Nelson told us, "Walter would not confirm anything to me. I know that he was there, but he wouldn't admit it, even to me."

5. Walter Haut, personal interviews, 1989-2002.

CHAPTER 25

1. Doleman, *Archeological Testing.*

2. Ibid.

3. See complete report by Anthony Bragalia in Appendix IV.

4. Anthony Bragalia, personal email to Tom Carey, October 17, 2008.

POSTSCRIPT

1. Carey, Thomas J., and Donald R. Schmitt. *Roswell: The Ultimate Cold Case CLOSED*. (Newburyport, MA: New Page Books/Red Wheel Weiser, 2020).

2. Bragalia, Anthony. "Renowned U.S. Navy Commander Reveals Stunning Roswell Crash Secret." *UFO Explorations*, 2010,. See also, Birnes, William J., and Joel Martin. *UFOs and the White House*. (New York: Skyhorse Publishing, 2018).

3. Carey and Schmitt, 2020.

BIBLIOGRAPHY

Anderson, Jack, and Michael Binstein. "Air Force Tried to Lead Everyone Astray on Roswell Incident." *Albuquerque Journal,* June 1, 1995.

Atomic Blast (RAAP base newspaper). Public Information Office of the 509th Bomb Group, RAAP.

Berlitz, Charles, and William L. Moore. *The Roswell Incident.* New York: Grosset & Dunlap, 1980.

Birnes, William J. "Squiggly." *UFO Magazine* 21, No. 4, June 2004.

Bragalia, Anthony. "Science Reports Show Roswell Crash Was ET." Previously unpublished.

Broad, William J. "Wreckage of a 'Spaceship': Of This Earth (and U.S.)." *The New York Times,* Sept. 18, 1994.

Brokaw, Tom. *The Greatest Generation.* New York: Random House, 1998. Bullard, Thomas E. "Folklore Scholarship and UFO Reality." *International UFO Reporter,* September/October 1988.

Campbell, John W., Jr. "Who Goes There?" *Astounding Science Fiction.* New York: Street & Smith, August 1938.

Carey, Thomas J. "Will the Real Sheridan Cavitt Please Stand Up?" *International UFO Reporter* 23, No. 3, Fall 1998.

Carey, Thomas J., and Donald R. Schmitt. *Witness to Roswell: Unmasking the 60-Year Cover-Up.* Franklin Lakes, NJ: New Page Books, 2007.

———. *Witness to Roswell* [magazine]. Roswell, NM: Triton, 2003.

———. in *The Roswell Dig Diaries,* a SciFi Channel Book, edited by Mike McAvennie, New York: Pocket Books, 2004.

Dennis, Glenn. *1947 Roswell Incident.* Roswell, NM: Alpha-Omega, 1991.

Doleman, William H. *Archeological Testing and Geophysical Prospecting at the Reported Foster Ranch UFO Impact Site, Lincoln County, New Mexico.* Albuquerque, NM: Office of Contract Archeology, University of New Mexico, 2003.

Fuhrman, Mark. *Murder in Greenwich: Who Killed Martha Moxley?* New York: HarperCollins, 1998.

"General Ramey Empties Roswell Saucer." *Roswell Daily Record,* July 9, 1947. Hastings, Robert L. *UFOs and Nukes: Extraordinary Encounters at Nuclear Weapons Sites.* Albuquerque, NM: Self-published, 2007. *ufohastings.com.*

"Harassed Rancher Who Located 'Saucer' Sorry He Told about It." *Roswell Daily Record,* July 9, 1947.

Headquarters United States Air Force. Captain James McAndrew. *The Roswell Report: Case Closed.* Washington, DC: Barnes & Noble, by arrangement with the U.S. Government Printing Office, 1997.

———. Colonel Richard L. Weaver. *Report of Air Force Research regarding the "Roswell Incident."* Washington, DC: U.S. Government Printing Office, 1994.

———. Colonel Richard L. Weaver and First Lieutenant James McAndrew. *The Roswell Report: Fact vs. Fiction in the New Mexico Desert.* Washington, DC: U.S. Government Printing Office, 1995.

Lawrence, Jill H. "Did a UFO Really Crash at Roswell?" *Northeast Ohio Paradigm News,* September 1996.

LeMay, John. *Images of America: Roswell.* Charleston, SC: Arcadia Publishing, 2008.

McAvennie, Mike, ed., with contributions from William H. Doleman, PhD, Thomas J. Carey, and Donald R. Schmitt. *The Roswell Dig Diaries: A Sci Fi Channel Book.* New York: Pocket Books, 2004.

Moore, William L. "The Roswell Investigation: Update and Conclusions." Prescott, AZ: Self-published, 1981.

Morris, Robert. "UFO whistleblower's nephew maintains family tradition of belief in aliens." *Houma Today,* October 29, 2007.

"New Base Surgeon Assigned to Sqdn. M." *The Atomic Blast.* Public Information Office of the 509th Bomb Group, RAAP, February 21, 1947.

Pflock, Karl T. *Roswell: Inconvenient Facts and the Will to Believe.* Amherst, NY: Prometheus Books, 2001.

"RAAF Captures Flying Saucer on Ranch in Roswell Region." *Roswell Daily Record,* July 8, 1947.

RAAF Roswell, New Mexico. (509th Bomb Group/Roswell Army Air Field Yearbook.) Roswell, NM: RAAP, 1947.

Randle, Kevin D., and Donald R. Schmitt. *The Truth about the UFO Crash at Roswell.* New York: Avon Books, 1994.

————. *UFO Crash at Roswell.* New York: Avon Books, 1991.

Recollections of Roswell, Part II. Mount Rainier, MD: Fund for UFO Research, 1992.

Research Division, College of Engineering, NYU. *Technical Report No. 93.02: Constant Level Balloons,* January 31, 1949.

Rice, Patricia D. "Roswell Aliens: Air Force Deception." Letter to the *Dallas Morning News,* July 6, 1997.

Richelson, Jeffrey. *The U.S. Intelligence Community: Organization, Operations, and Management, 1947-1989,* 3rd Edition. Boulder, CO: Westview Press, 1995.

The Roswell Crash: Startling New Evidence. The SciFi Channel, 2002.

Roswell Remembered: A City on the Edge of History. Crystal Sky Productions, 1995.

Roswell: The UFO Cover-Up. Showtime/Viacom, 1994.

Ruppelt, Edward J. *The Report on Unidentified Objects.* Garden City, NY: Doubleday, 1956.

Schiff, Steven. "News Release." Release of GAO report detailing results of records search related to events surrounding alleged 1947 UFO crash near Roswell, NM, July 28, 1995.

Shawcross, Tim. *The Roswell File.* Osceola, WI: Motorbooks, 1997.

Shirkey, Robert J. *Roswell 1947: "I Was There."* Roswell, NM: Movin' On, 1999.

Stowers, Carlton. "A half century later, witnesses insist little green—or maybe brown—men crashed in New Mexico." From *dallasobserver.com*. Originally published by the *Dallas Observer*, April 3, 2003.

Stringfield, Leonard H. "Roswell & the X-15: UFO Basics." *Mufon UFO Journal* No. 259, November 1989.

———. "We Are Now a Part of History." *UFO Crash/Retrievals Status Report VI: The Inner Sanctum*. Cincinnati, OH: Leonard H. Stringfield, 1991.

———. *The UFO Crash/Retrieval Syndrome*. Seguin, TX: MUFON, 1980.

Thurber, John. "Allan Grant, 88; shot iconic photos for *Life* magazine." *Los Angeles Times* obituaries, February 9, 2008.

Tilley, John M., and Larry Tilley. *Exposé Roswell UFO Incident (Corona Crash)*. Roswell, NM: Lulu, 2007.

UFO Secret: The Roswell Crash Update 2000. Venice, CA: UFO Central Home Video, 2001.

"UFOs: Seeing Is Believing." *Peter Jennings Reporting*. ABC News, February 24, 2005.

United States General Accounting Office. *Results of a Search for Records Concerning the 1947 Crash near Roswell, New Mexico*. Pub. No. GAO/NSIAD-95-187. Washington, DC: GAO, 1995.

Vigil, Maurilio, and Roy Lujan. "Parallels in the Careers of Two Hispanic U.S. Senators." *New Mexico Historical Review* No. 47, October 1972.

Weaver, Richard L. *The Roswell Report: Fact vs. Fiction in the New Mexico Desert*. U.S. Air Force, 1995.

Weisgall, Jonathan M. *Operation Crossroads: The Atomic Tests at Bikini Atoll*. Annapolis, MD: Naval Institute Press, April 1994.

INDEX

U

UFO Crash at Roswell (Randle and
   Schmitt), 77, 217
Unsolved Mysteries (TV show),
   8, 222
"Up an' Atom" (B-29), 106–107
U.S. Army Air Forces. *see also*
   Roswell Army Air Field (RAAF)
   dummy explanation, 8–9
   flying saucer explanation, 7
   Project Mogul explanation,
      8–9, 14, 15
   weather balloon explanation,
      7, 8

V

van Buskirk, Adrian, 250
Van Why, Raymond, 126, 127
Vega, Frank, 188–189, 190
Vegh, Elaine, 71
von Braun, Wernher, 251, 252

W

Wahnee, Meyers, 223–224
Wallace, LeRoy, 128
Warne, Harold M., 117
Warnke, William, 130
Washington Post, 86
weather balloon explanation, 7, 8,
   85–88

Weaver, Richard L., 20–23
Weinbrenner, George, 150
Wells, Jeff, 215–216
Whalen, Steve, 103–104
Whitmore, Walt, Sr., 44
   attempt to hide/interview
      Brazel, 55
   threats against, 174–175
Wilcox, George M., 12, 18, 85,
   88, 223
   meeting with Brazel, 41, 54
   threat to Anaya, 182–183
   threat to Dennis, 182
   warning to Anayas, 81
Wilcox, Inez, 223
Wiley, James R., 103
Wilmot, Dan, 13
Wirth, Joseph T., 28–29
witness case, Roswell Incident as,
   14–15
witnesses. *see also* alien bodies
   of alien bodies, 40, 71, 72,
      79–80, 127
   Barton, Chester P., 95–96
   Benavidez, Eleazar, 139–143
   Benthal, Frederick, 128–130
   Bush, Miriam "Andrea,"
      117–121
   Dennis, Glenn, 130–133, 134
   of eighteen-wheel lowboy
      cargo, 90–92, 94–95,
      97–98
   Ellingsworth, Leo, 130

ABOUT THE AUTHORS

Thomas J. Carey, a native Philadelphian, holds degrees from Temple University (BS in Business Administration) and California State University, Sacramento (MA in Anthropology), and also attended the University of Toronto's PhD Programme in Anthropology.

An Air Force veteran who held a TOP SECRET/CRYPTO clearance, Tom is now a retired Philadelphia-area businessman. He was a Mutual UFO Network (MUFON) State Section Director for Southeastern Pennsylvania from 1986 to 2001, a Special Investigator for the J. Allen Hynek Center for UFO Studies (CUFOS) from 1991 to 2001, and a member of the CUFOS board of directors from 1997 to 2001.

Tom has authored or coauthored numerous published articles about the Roswell events of 1947 and has contributed to a number of books on the subject as well. He has appeared as a guest on many radio and TV shows throughout the country, including *Coast to Coast AM* with Art Bell and George Noory, *Fox and Friends, Comcast Network Friends,* and *Larry King Live!* and has contributed to a number of Roswell-related documentaries, on-screen and behind the scenes. Tom was a consultant and interviewee on the highly acclaimed and rated 2002, two-hour SyFy Channel documentary, *The Roswell Crash: Startling New Evidence,* the History Channel's *Conspiracy Theory,* the Travel Channel's *Weird Travels: Roswell,* the SyFy Channel's *Sci Fi Investigates,* and the History Channel's *UFO Hunters.* He was also a featured interviewee on the 20th Century Fox Golden Anniversary re-release DVD of the 1951 sci-fi classic, *The Day the Earth Stood Still.* Most recently, Tom was the historical consultant for the 2021 Discovery+ Channel's six-part series on the 1947 Roswell crash.

The first and second editions of *Witness to Roswell: Unmasking the 60-Year Cover-Up* were the #1 best-selling UFO books in the world for three years (2007-2009). A book about the secret history of Wright-Patterson Air Force Base in Dayton, Ohio, titled *Inside the Real Area 51*, also co-authored with Don Schmitt, reached #1 in Amazon's "Astronomy & Space Science" category and was also in its "Top 100 Books of the Year" for 2013. Another book about Roswell, *The Children of Roswell*, was released in 2016 and also reached #1 in its category on Amazon. *UFO Secrets inside Wright-Patterson*, released in 2019, remains a best seller; and *Roswell: The Chrono-logical Pictorial* and *Roswell: The Ultimate Cold Case CLOSED* were both released in 2020. *Touched by Roswell: Crash Encounters of the Rich and Famous* was released in 2021.

Tom and his wife Doreen have two grown children and reside in Huntingdon Valley, Pennsylvania.

Donald R. Schmitt is a graduate of Concordia University and a former codirector of the J. Allen Hynek Center for UFO Studies (CUFOS) in Chicago. For eight years he was a special investigator for Hynek who is still considered the foremost authority on the subject of UFOs. For the next ten years he would serve as director of special investigations at CUFOS and was also art director for the *International UFO Reporter*.

A seven-time best-selling author with a total of seventeen books, Schmitt's first book *UFO Crash at Roswell*, coauthored with Kevin D. Randle, was made into the Showtime Golden Globe nominated "Best Picture Made for TV" award for the movie ROSWELL. He is also the cofounder of the world-famous International UFO Museum and Research Center (IUFO-MRC) in Roswell, New Mexico, where he serves as advisor and honorary board member for life. Don is also the North American representative to the International Coalition on Extraterrestrial Research (ICER) which presently includes twenty-eight countries with the goal of United Nations recognition. An international lecturer and movie, TV, and features consultant on the subject, he has appeared in over thirty documentaries and has guested on *Oprah*,

48 Hours, Larry King Live, Good Morning America, FOX Nation, Peter Jennings Reports, Paul Harvey News, and hundreds more. Don has lectured on Roswell in forty-eight states, Canada, Mexico, South America, Europe, the UK, Russia, China, Japan, and Australia. Don and his wife Marie reside on a ranch in southeastern Wisconsin.*Straight Flush* (B-29) *continued*